MATHS IN ACTION

National 4
Lifeskills

R Howat
J McLaughlin
G Meikle
J Morgan
E Mullan
D Murray
R Murray
J Pritchard
E Thomson

Series editor: E Mullan

Nelson Thornes

Published in 2013 by:
Nelson Thornes Ltd
Delta Place
27 Bath Road
CHELTENHAM
GL53 7TH
United Kingdom

13 14 15 16 17 / 10 9 8 7 6 5 4 3 2 1

A catalogue record for this book is available from the British Library

ISBN 978 1 4085 1911 0

Cover photograph: Rubberball/Mike Kemp/Getty
Page make-up and illustrations by Tech-Set Ltd, Gateshead

Printed and bound in Spain by GraphyCems

Contents

Introduction

This book has been created especially to cater for the needs of the student attempting the National 4 Lifeskills course. It is important that a student sees the relevance of a subject in order to learn it effectively.

With that in mind, wherever possible, real-life scenarios from daily living, the workplace, and a variety of school subjects have been chosen to give questions a context. Inevitably, these contexts do get simplified as other factors not relevant to the occasion are omitted.

Each chapter is structured in the same way, and begins with a problem associated with a real-life situation. Although some students may be able to attempt this already, it will prove to be a challenge to most, as they are unlikely to have the necessary skills and experience to solve the problem fully. It should therefore act as an introductory stimulus for bringing into play the strategies for problem solving that students are already familiar with, while perhaps also showing the shortcomings of these for the task in hand. By the time students have worked through the chapter, they should be able to complete the problem successfully, and when the problem is reintroduced at the end of the chapter, they should appreciate how their skills have expanded.

There follows a feature called 'What you need to know', which contains a few questions that students should be able to tackle using their current knowledge. It is important that students are comfortable with each of these, because the questions exercise the very skills required to get the most from the chapter.

Exposition of the new skills is accompanied by worked examples, which the student is encouraged to mimic in the laying out of their own solutions.

The level of difficulty of questions is indicated in the following ways:

Questions essential to the successful completion of the course are contained in the A exercises. If students can do these, they are on target for success.

A student wishing to take Maths further into National 5/National 5 Lifeskills should also be able to handle the strategies needed to complete the B exercises.

B exercise questions that have a green-tinted background are more challenging, and are suitable for those who want to be stretched further.

At the end of each chapter there is a section called 'Preparation for assessment'. If a student cannot do one of these questions they should perhaps revisit the corresponding section in the chapter or ask the teacher for help. This section culminates with a revisit to the introductory question, which by now should be totally accessible. This will act to bring the class together again if students have been working at different levels.

Throughout the book, various icons have been used to identify particular features:

  indicate the most appropriate mode of tackling a problem. As part of the assessment measures students' numerate abilities without the aid of a calculator, it is essential that the parts marked 'non-calculator' are indeed done without a calculator.

 indicates a topic suitable for class discussion. This device is used when, as a topic starts, we are aware that students may have different educational experiences as they advanced through S1 to S3 following the CfE. A class discussion will help to make the necessary knowledge base the same for the whole group.

 indicates a question for which some research or investigative work would enrich the student's experience.

 indicates where a puzzle has been added also for the sake of enrichment. These puzzles can always be taken further.

Opportunities exist, especially in the area of finance, for exploiting the power of the spreadsheet. The use of IT should be thoroughly explored.

Statistics is to be seen as a practical subject, and, as well as the exercises provided, investigative surveys leading to a written report would be beneficial. This too is made more pertinent and enjoyable through the use of IT.

Finally, a convention is followed throughout the series whereby the decimal point is placed mid-line, e.g. when writing '3 point 14' we would write 3·14 and not 3.14. The latter style is used in spreadsheets, however.

In Britain we often use the 'point on the line' to act as a less conspicuous multiplication sign than '×'. This is especially useful in algebra:

e.g. When $x = 6$ then $3x + 2 = 3.6 + 2 = 18 + 2 = 20$

In conclusion, maths is everywhere, it's relevant, it *is* essential, and it can be enjoyed.

1 Earnings

⏸ Before we start...

Jess is on work experience at a local accountancy firm.

She is given some specimen monthly pay slips to complete.

Here is one of them:

Payment details	Units worked (hours)	Rate (£/h)	Amount (£)
Pay	23·00	9·75	

Deductions	PAYE tax (£)	National Insurance (£)	Total deductions (£)
	44·85	13·46	
		Net pay	

The yellow rectangles have been left blank for Jess to fill in.

She is unsure of the meaning of some of the terms and has to ask for help ...

▶ What you need to know

1 Calculate:

   a 6% of £240   **b** 7·5% of £3284.

2 Express as a fraction of an hour:

   a 15 minutes   **b** 45 minutes   **c** 12 minutes

   d 40 minutes   **e** 10 minutes   **f** 25 minutes.

3 Express the following in minutes:

   a 0·3 hour   **b** 0·1 hour   **c** 0·8 hour   **d** 1·4 hours

   e 12·7 hours   **f** $5\frac{2}{3}$ hours   **g** $2\frac{5}{6}$ hours.

1.1 Hourly rate and overtime

Many jobs describe their pay by quoting an hourly rate ... the amount you're paid for every hour of work.

Overtime is paid for work carried out at 'unsociable times', maybe at weekends or at night.

The most common rates of overtime are:

Double time: 2 × rate of pay ... you're paid for 2 hours for every hour worked.

Time-and-a-half: 1·5 × rate of pay ... you're paid for 1·5 hours for every hour worked.

Example 1

Samuel did a 7-hour shift at the factory. His rate of pay is £10·40 per hour.

Calculate what he earned for the shift.

7 × 10·40 = £72·80

Example 2

Jack fills shelves at a supermarket.

For working on Sunday nights he is paid time-and-a-half.

Calculate his pay one Sunday night when he works for $3\frac{1}{2}$ hours and his basic rate of pay is £8·40 an hour.

3·5 × 1·5 × 8·40 = £44·10

Exercise 1.1A

1 Calculate the pay earned for:

    a 8 hours at £11·35 per hour   **b** $6\frac{1}{2}$ hours at £7·52 per hour.

2 A senior staff nurse position is advertised online.

    How much is the weekly pay for this position?

> BASE PAY: £13·22/HOUR
> OTHER PAY: COMPETITIVE
> Senior Staff Nurse
> CONTRACTED HOURS: 33 hrs/wk
>
> Apply Now

3 Jody is studying to be a primary school teacher.

    She works in a high street store during the holidays and at weekends to help pay for her studies.

    She has a contract for 8 hours per week at £7·04 per hour with an option of extra work if it is available. There is no overtime rate.

    One week she works an extra 3 hours.
    How much is she paid that week?

4. A local paper advertised this job. Martin applied for and got the job.

In his first week, Martin worked 2 hours on Saturday and $2\frac{1}{2}$ hours on Sunday.

How much was he paid for his overtime work?

HOSPITAL CLEANER REQUIRED

HOURS TO BE NEGOTIATED
BASIC RATE OF PAY: £9·15/HOUR

OVERTIME RATES
SATURDAY: TIME-AND-A-HALF
SUNDAY: DOUBLE TIME

5. Kerr earns a total of £363·20 for a 32-hour a week job in a precision engineering factory.

What is his hourly rate of pay?

6.

Clerical Assistant and Receptionist

Job description   Details

Job Title: Apprentice Clerical Assistant and Receptionist
Salary: £15,652 per annum for 35 hours per week

This job is advertised on an internet site. Assume 52 weeks in a year.

How much is the hourly rate?

Exercise 1.1B

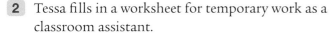

1. Steven works for a local council parks department.

His rates of pay are as follows:

Basic rate: £11·26 per hour

Overtime 1: 5 p.m. until midnight – time-and-a-half

Overtime 2: midnight until 5 a.m. – double time.

One night Steven grits the icy roads.
He works from 7 p.m. until 1 a.m.

How much is he paid for this work?

2. Tessa fills in a worksheet for temporary work as a classroom assistant.

Parts of a day are written as fractions of a day.

Her rate of pay is £38 per full day of work.

   a Calculate the total time, in days, worked that week.

   b Calculate how much Tessa is paid for her week of work.

Week ending	06/12/13
Day	Fraction worked
Monday	1
Tuesday	$\frac{3}{4}$
Wednesday	$\frac{1}{2}$
Thursday	$\frac{3}{4}$
Friday	$\frac{1}{2}$

3 Liam divides his time working at a swimming pool by giving swimming lessons and being a lifeguard.

His pay for giving lessons is £11·36 per hour.

His pay for being a lifeguard is £7·15 per hour.

One week he gives $4\frac{1}{2}$ hours of lessons and 6 hours acting as a lifeguard.

What are his earnings this week?

4 A worker earns £189·80 for a week's work, which includes some overtime at double time.

The basic rate of pay is £7·30 per hour.

If 20 hours are worked at the basic rate, how many hours are worked at double time rate?

5 Max works for a letterbox marketing company delivering free newspapers in a residential area of Edinburgh.

He is paid £7·50 for every 80 newspapers he delivers.

He is also given £6 per day for travelling expenses.

In 4 days he delivers 1040 newspapers.

How much is he paid altogether?

1.2 Performance related pay

Performance related pay is designed to stimulate greater output from workers.

Under this heading come bonus, commission, and piecework.

- A **bonus** is an extra payment made by the employer if certain targets are met.
- On top of his wage, a salesman will often be paid a percentage of the value of the sales he makes: this is known as **commission**.
- When a worker is making items, she is often paid according to the number of items she makes. This is referred to as **piecework**.

Example 1

A car salesman is on 2% commission.

This means he will earn 2% of the selling price of any car he sells.

What would be the commission earned on a car sold for £10 590?

1% of 10 590 = 105·90

so 2% of 10 590 = 211·80

So the commission earned is £211·80.

Example 2

Angela works in an electronics factory.

She earns a basic wage for making 25 circuit boards a day.

She has the chance to make extra money.

For every circuit board over 25 completed in a day, she is paid £6·50 per board.

How much extra pay does Angela earn on a day that she completes 29 boards?

$29 - 25 = 4$ extra boards.

$4 \times 6·50 = 26$

So Angela earns an extra £26.

Exercise 1.2A

1   An advertisement appeared in a newspaper:

RASPBERRY PICKERS REQUIRED
RATE OF PAY:
PIECEWORK
THE MORE YOU PICK
THE MORE YOU EARN!
£0·35 per filled punnet

   a   Rachael picks raspberries as a summer holiday job.
       She is paid daily.
       How much is she paid for filling 76 punnets?
   b   Rachael's friend is paid £28 one day.
       How many punnets had he filled?

2   In a woollen mill, employees are paid by piecework.

	Rate of pay
Cuff linking	£0·80 for 10
Collar linking	£1·30 for 10
Seam linking	£1·60 for 10

   a   Mary links 1800 cuffs in a week.
       How much is she paid for her week's work?
   b   Jane links seams. Here is a record of her work one week.

Monday	Tuesday	Wednesday	Thursday	Friday
300	280	320	270	300

       i   What is the total number of seams that she linked?
       ii How much is she paid for her week's work?

3   As an incentive to its sales staff, a city garage gives a commission of
    1·2% of the selling price on all car sales.
    How much commission is made on a car selling for £13 250?

4   An agricultural engineering company makes grain
    dryers for farms.

    The sales staff are paid commission of 0·9% of the
    sale price of a dryer.

    What commission is made on a dryer that costs
    £18 000?

5 Gordon works off-shore as a welder on the oil rigs in the North Sea.

He is paid £900 per week.

After working for a year, he receives a bonus of an extra month's pay.

    a How much is his bonus if we assume that a month has 4 weeks?

    b After a pay rise, his annual income is £48 000.

       He works for a year with this income.

       How much will he get for his new bonus?

6 Shaun works as a plasterer for a building company.

His rate of pay is £17·48 per hour for a $7\frac{1}{2}$ hour day.

He is given a contract, which he must complete in 21 days.

If he completes the contract in *less* than 21 days, he will receive a bonus of one day's pay for every day under the 21.

If Shaun completes the contract in 19 days, how much is his bonus?

Exercise 1.2B

1 Autocars pay their sales people 1% of all sales as commission.

Fraters pay their sales people 2% on sales over £5000.

Who pays more commission for a car with a selling price of £12 900?

2 Harry is a collar linker in a woollen mill. He is paid by piecework.

1 to 45 collars: £1·76 per collar

More than 45 collars: £2·15 per collar over the 45

How much pay does Harry earn one day when he links 53 collars?

3 Margot has a permanent part-time job in a city store. The hours she's paid for are:

Thursday: 4 p.m. to 9 p.m.

Saturday: 9 a.m. to 5 p.m.

Sunday: 11 a.m. to 3 p.m.

Her rate of pay is £8·70 per hour.

On Sundays she is paid time-and-a-half.

    a **i** How much does she earn in a week?   **ii** How much does she earn in a year?

    b Margot gets an end of year bonus. The bonus is 14% of her earnings for a year.

       How much is her bonus?

4 To increase sales, till staff are encouraged to 'add to the basket' by asking customers if they would like to buy the 'product of the week'.

Product of the week
Pack of 3 microfibre cloths
£2

The following table shows sales information for one day.

	Kirsty	Beccy	Alan	Fred
Number of transactions	240	225	180	260
Number of sales of 'product of the week'	24	9	9	39

As a bonus, whoever has the highest percentage sales of 'product of the week' receives a £15 in-store gift card.

Who wins the store gift card?

1.3 Deductions – income tax

The largest deduction from earnings is **income tax**.

This is a compulsory levy on earnings by the government to pay for social security, defence and law enforcement.

Income tax was introduced in Britain by William Pitt the Younger, the Prime Minister, in 1799.

The incentive for such a tax was to pay for weapons and equipment in preparation for the war against Napoleon Bonaparte (the Napoleonic Wars).

You can explore these historical facts using the internet.

The picture shows a statue of Pitt in George Street, Edinburgh.

Example 1

In March 2012, the Chancellor of the Exchequer set the following rates for income tax in the year 2012 to 2013:

    Personal allowance = £8105

    Basic rate of income tax = 20%

Calculate the yearly amount of tax deducted from an annual pay of £22 160.

Annual pay = £22 160.

Personal allowance = £8105.

Taxable income = £22 160 − £8105 = £14 055.

Income tax = 20% of taxable income
        = 0·2 × 14 055
        = £2811

Example 2

Calculate the monthly amount of income tax paid on an annual pay of £25 600, given that the personal allowance is £8105 and the basic rate of income tax is 20%.

Annual pay = £25 600.

Personal allowance = £8105.

Taxable income = £25 600 − £8105 = £17 495.

Income tax = 20% of taxable income

$\quad\quad\quad\quad = 0{\cdot}2 \times £17\,495$

$\quad\quad\quad\quad = £3499$

Monthly tax paid = £3499 ÷ 12 = £291·58.

Exercise 1.3A

Throughout this exercise, personal allowance = £8105, rate of income tax = 20%.

1 For each of the following annual incomes, calculate:

   i the taxable income

   ii the amount of tax paid.

   a £32 000   b £19 450   c £29 475.

2 Jill works as a nurse at a general hospital.
   She earns £1086 per month.

   a How much does Jill earn in a year?

   b How much is her taxable income?

   c How much tax does Jill pay in a year?

3 Mark's annual pay is £20 085.
   Calculate:

   a his taxable income

   b the tax he pays annually

   c his monthly tax bill, assuming he pays by calendar month.

4 Roger gets a pay rise from £25 700 to £27 300 per annum.

   a How much income tax does Roger pay before the pay rise?

   b How much *more* tax does Roger pay after the pay rise?

5 Annie joins a company and earns £18 200 per year.
   After working for the company for a year, she is promised a pay rise of 2·5%.

   a How much is her annual pay after the pay rise?

   b How much income tax does Annie pay after the pay rise?

Budget changes

Rates of income tax and personal allowances can vary from year to year.

These changes are announced when the Chancellor of the Exchequer makes his annual speech, the Budget.

The word 'Budget' is derived from a French word, *bougette*, meaning 'little bag'.

Traditionally the Chancellor of the Exchequer carries his speech to Parliament in a little red box.

Exercise 1.3B

1   The table shows tax details for two different years.

Tax year	A	B
Rate of tax	21%	23%
Personal allowance	£4980	£5430

    a   Calculate, for both years, on an income of £25 600:
       i the taxable income   **ii** the amount of tax paid annually.

    b   In which year is more tax paid?

2   Hugh was a tanker driver when the rate of tax was 22%.

His annual income was £23 100.

His personal allowance was £5860.

Calculate:

    a   his taxable income

    b   the tax paid annually

    c   the tax paid monthly.

3   Stuart works on the quayside. The basic tax rate is 24%.

His personal allowance is £5375 and his taxable income is £17 500.

    a   Calculate:
       i his annual pay
      ii his annual tax bill
      iii his monthly tax bill.

    b   Helen also works on the quay. She pays £3840 in tax annually.

       Can you work out what her taxable income is?

1.4 Deductions – National Insurance

The second largest deduction from pay is **National Insurance**.

This is a system of compulsory contributions from employees and employers. You and your employer pay into the scheme and it provides assistance during sickness, unemployment and retirement.

Find out more about National Insurance (NI) on the internet.

- When was NI first introduced in Britain?
- Who was responsible for introducing it?
- Why was it introduced?

Employees' contribution

What you pay into the scheme depends on what you earn.

Your earnings	National Insurance	Formula
Up to £7592	0%	No NI contribution payable
Between £7592 and £42 484	12%	$= (\text{pay} - 7592) \times 0.12$

The above rates of NI for employees are for the tax year 2012 to 13, as set out in the Budget speech in March 2012 by the Chancellor of the Exchequer.

Example 1

Ruaridh is a steeplejack. He earns £16 380 per annum.

How much National Insurance does Ruaridh pay?

£16 380 – £7592 = £8788

12% of £8788 = 0.12 × 8788
            = £1054.56

Ruaridh pays £1054.56 National Insurance in a year.

Example 2

Bill works in a bank. His monthly income is £3240.

How much National Insurance does Bill pay:

a in a year **b** in a month? (Use the rates given in the table above.)

a Annual income = £3240 × 12 = £38 880.

National insurance is due on £38 880 − £7592 = £31 288.

12% of £31 288 = 0.12 × 31 288
              = 3754.56

Bill pays £3754.56 National Insurance in a year.

b Bill pays £3754.56 ÷ 12 = £312.88 National Insurance in a month.

Exercise 1.4A

Throughout this exercise use the rates of National Insurance set out in the table on page 10.

1 Fiona is a civil engineer with the roads department.

She has an annual pay of £29 120.

How much National Insurance does she pay in a year?

2 Chas has a monthly pay of £3330.

   a How much does he earn in a year?

   b How much National Insurance does he pay in a year?

   c How much National Insurance does he pay in a month?

3 A company uses a spreadsheet to calculate the National Insurance contributions of its employees.

Cell C2 holds the formula: =(B2–7592)*0.12

The value in the cell is 1752·96.

   a **i** What formula is there in cell C3?

      ii What value will appear in the cell?

   b Calculate the value in cell:

     i C4   **ii** C5.

   c Cell D2 has calculated the monthly deduction for J Brown.

     What formula is in D2?

	A	B	C	D	E
	Name	Annual pay (£)	Annual N.I. employee contribution	Monthly deductions	
1					
2	J Brown	22200	1752.96	146.08	
3	O Smith	26040	A	D	
4	B Thomas	21480	B	E	
5	S Green	30100	C	F	
6					
7					
8					

   d Calculate the monthly deduction for: **i** B Thomas   **ii** S Green.

4 Employers *also* make National Insurance contributions on behalf of employees.

The table shows how they calculate their contribution.

Your earnings	National Insurance	Formula
Up to £7592	0%	No NI contribution payable
Between £7592 and £42 484	13·8%	= (pay − 7592) × 0·138

Column E of this spreadsheet calculates the employer contribution.

	A	B	C	D	E	F
	Name	Annual pay (£)	Annual N.I. employee contribution	Monthly deductions	Employer annual contribution	
2	J Brown	22200	1752.96	146.08	2015.90	
3	O Smith	26040	A	D	G	
4	B Thomas	21480	B	E	H	
5	S Green	30100	C	F	I	

a What formula is contained in: **i** E2 **ii** E4?

b Calculate the values in: **i** E3 **ii** E4.

c Counting both sets of contributions, how much was paid into the scheme for S Green?

For large earners (people earning more than £42 484) an extra rate of National Insurance contributions is added.

Your earnings	National Insurance	Formula
Up to £7592	0%	No NI contribution payable
Between £7592 and £42 484	12%	= (42 484 − 7592) × 0·12
Greater than £42 484	2%	= (pay − 42 484) × 0·02

Example 3

Clive is an air traffic controller and earns £55 000 a year.

Calculate how much he pays in National Insurance contributions.

Contribution at 12%: (42 484 − 7592) × 0·12
= 34 892 × 0·12 = £4187·04

Contribution at 2%: (55 000 − 42 484) × 0·02
= £250·32

Total NI contributions made by Clive
= 4187·04 + 250·32 = £4437·36.

Exercise 1.4B

1 A member of the Scottish Parliament gets a salary of £53 091.

   a How much over the £42 484 threshold is this?

   b How much should the National Insurance contribution be to cover the £42 484?

   c How much should be paid on the extra?

   d What total NI contribution should be made?

2 Gordon Black is a banker earning £68 000 per annum.

   a Calculate the amount of National Insurance he has to contribute.

   b Gordon Black's employer also had to pay National Insurance on his behalf.

Your earnings	National Insurance	Formula
Up to £7592	0%	No NI contribution payable
Between £7592 and £42 484	13·8%	= (42 484 − 7592) × 0·138
Greater than £42 484	2%	= (pay − 42 484) × 0·02

How much National Insurance was paid by Gordon's employer on his behalf?

3 A member of the UK Parliament earns £65 738. Calculate his National Insurance contributions.

1.5 Gross pay and net pay

Gross pay, sometimes called 'top line pay', is pay *before* deductions are made.

Net pay, sometimes called 'take home pay', is the *actual* pay received.

Pay month	February		Name	E Pearson	

Pay (£)		Deductions (£)			
Basic	1600·00	Tax	257·59	Gross pay	1600·00
Overtime	–	NI	102·67	Deductions	360·26
Total pay	1600·00	Total deductions	360·26	Net pay	1239·74

In this pay slip we see:

Gross monthly pay is £1600·00.

Total deductions are £360·26.

Net monthly pay is £1239·74.

If you're employed, a system called Pay As You Earn (PAYE) is used by your employer to deduct income tax and National Insurance contributions from your wages before you are paid.

Example 1

Andy Beattie is given an annual summary of his earnings.

Annual Summary Name A Beattie

Pay (£)		Deductions (£)			
Basic	28 420	Tax	4389·00	Gross pay	28 420
Overtime	–	NI	5098·00	Deductions	A
Total pay	28 420	Total deductions	A	Net pay	B

Calculate what should be entered in the cell labelled: **a** A **b** B.

a Cell A: Total deductions = £5098 + £4389 = £9487.

b Cell B: Net pay = £28 420 − £9487 = £18 933.

Exercise 1.5A

1 Here is K Knight's pay slip:

Pay month December   **Name** K Knight

Pay (£)		Deductions (£)			
Basic	1820·50	Tax	286·78	Gross pay	A
Overtime	–	NI	129·02	Deductions	B
Total pay	A	Total deductions	B	Net pay	C

What value should appear in cell:

a A   **b** B   **c** C?

2 Karen is a physiotherapist in a hospital.

Complete her pay slip.

Pay month	December		Name	Karen Duncan	

Pay (£)

Basic	2130·70
Overtime	–
Total pay	A

Deductions (£)

Tax	315·44
NI	163·01
Total deductions	B

Gross pay	A
Deductions	B
Net pay	C

3 Dawn has a part-time job as a nurse working in an Accident and Emergency Department.

She receives her monthly pay slip online.

Employee name		Date	National Insurance number	
Dawn Murray		30/05/13	YM987654A	
Hours	**Rate (£/h)**	**Amount (£)**	**Deductions**	**Amount (£)**
40·5	9·36	A	PAYE (tax)	74·64
			NI	35·09
			Total deductions	B
			Net pay	C

Calculate:

a the amount in box A

b the amount in box B

c net pay (box C).

4 Will receives his pay cheque every fortnight for working in a supermarket.

His salary includes some overtime at double time.

Name	Date	National Insurance number
Will Beaton	15/03/14	YM876543A

Payment	Hours	Rate (£/h)	Amount	Deductions	Amount (£)
Wage	10	7·12	A	PAYE (tax)	25·63
Overtime	4	14·24	B	NI	0·00
Total gross			C	Total deductions	D
				Net pay	E

Calculate:

a the amount in:

   i box A   **ii** box B   **iii** box C   **iv** box D.

b Will's net pay (box E).

Exercise 1.5B

1 Complete this monthly pay slip for Neil Wood.

Name	Date	National Insurance number
Neil Wood	24/12/14	YM123456D

Payment	Hours	Rate (£/h)	Amount	Deductions	Amount (£)
Wage	35	7·12	A	PAYE (tax)	104·00
Overtime 1	6	time-and-a-half	B	NI	63·00
Overtime 2	3	double time	C	Total deductions	E
Total gross			D	Net pay	F

2 Jennifer and Robin work on the ferry.

Jennifer is paid a basic rate of £8·20 an hour.

a How much will she earn, gross, if she works 8 hours basic, 4 hours at time-and-a quarter and 6 hours at time-and-a-half?

b Robin gets paid at a basic rate of £7·50 per hour.

He works 9 hours at the basic rate.

How much has he got to work at time-and-a-half before he makes more than Jennifer?

3 Martin works in the maintenance department of the Council.

He works a basic 40-hour week at £8·50 an hour.

He is always sure of 4 hours' overtime at time-and-a-half and 2 hours at double time.

a Calculate his gross weekly wage given these hours.

b On the assumption that there are 50 working weeks in the year, use your answer to estimate his annual earnings.

c His personal allowance is £6475.
   i Estimate his taxable income.
   ii Work out his annual income tax bill, assuming it is charged at 20%.
   iii If tax is deducted weekly by PAYE, how much tax should be taken each week?

d Use his estimated annual earnings and the table below to help you estimate how much NI should be deducted each week.

Your earnings	National Insurance	Formula
Up to £7592	0%	No NI contribution payable
Between £7592 and £42 484	12%	= (pay − 7592) × 0·12

e Estimate Martin's weekly take-home pay.

Preparation for assessment

1 Annette sees this job advertised on an internet site.

How much is the weekly pay for this job?

> PAY: £10·36 per HOUR
> **GENERAL PRACTICE NURSE**
> WEEKLY HOURS: 34
>
> Further info ➡

2 Mary works in a department store.

The store offers its staff overtime for the following times:

Late opening Thursday: 7 p.m. till 9 p.m. – overtime rate time-and-a-half.

Sunday: 1 p.m. till 4 p.m. – overtime rate double time.

One week Mary worked overtime on both the Thursday and the Sunday.

Her basic rate of pay is £8·10 an hour. How much was her overtime pay that week?

3 A trained paramedic is paid £19 780 per annum.

He also gets a bonus of 15% of his annual income as an 'unsociable hour allowance'.

Calculate the total annual pay, which includes the bonus.

4   Sandra had an annual income of £21 075.

Her personal allowance was £8105.

Calculate:

a   her taxable income

b   her annual tax payment, given that tax was levied at 20%.

5   Archie is allowed to earn £7592 in a year before he starts to pay National Insurance. He pays 12% of his earnings over £7592.

How much National Insurance will he pay in a year with an annual income of £19 470?

6   Marsha is a gardener. Her pay slip for March is shown.

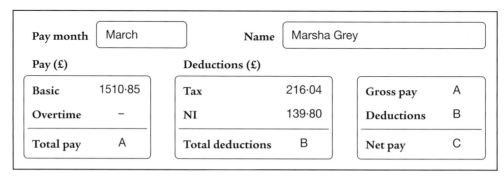

Pay month	March		Name	Marsha Grey	
Pay (£)		**Deductions (£)**			
Basic	1510·85	Tax	216·04	Gross pay	A
Overtime	–	NI	139·80	Deductions	B
Total pay	A	Total deductions	B	Net pay	C

What entry should there be in the cell labelled:

a   A   b   B   c   C?

7   Remember Jess? She is on work experience at a local accountancy firm.

She is given some specimen monthly pay slips to complete.

Here is one of them:

Payment details	Units worked (hours)	Rate (£/h)	Amount (£)
Pay	23·00	9·75	

Deductions	PAYE tax (£)	National Insurance (£)	Total deductions (£)
	44·85	13·46	
		Net pay	

The yellow rectangles have been left blank for Jess to fill in.

Find the values that should be in each yellow box.

❚❚ Before we start...

Rachel has 200 metres of temporary fencing to make a horse riding arena.

She wants the arena to have the largest possible area.

What shape would be best for the arena?

What would be the dimensions of the arena?

▶ What you need to know

1 Calculate the perimeter and area of:

a

2·1 m

1 m

b

2·3 m

2·3 m

2 Change:

    a 4·6 metres to centimetres   b 245 millimetres to centimetres

    c 93 centimetres to metres   d 1·78 metres to millimetres

    e 5602 metres to kilometres   f 2390 millimetres to metres.

3 The diameter of a colour wheel is 15·6 centimetres.

  What is the length of one blade of the wheel, i.e. its radius?

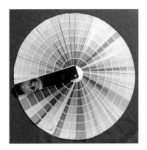

4 Round each measurement to 1 decimal place.

    a 3·14   b 124·79   c 78·65   d 35·95

5 Calculate the perimeter of each shape.

a

2·5 cm

6 cm

b

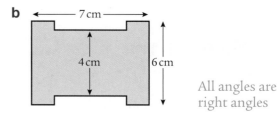

7 cm

4 cm

6 cm

All angles are
right angles

2.1 Tolerance

Fuzzy numbers

When we round off a number, we make it less accurate.

When we say that $\sqrt{2} = 1\cdot41$ correct to 2 decimal places, we are actually saying that
$\sqrt{2}$ lies between 1·405 and 1·415 ... any number, n, in the range $1\cdot405 \leqslant n < 1\cdot415$
will round to 1·41.

This fuzziness can be expressed as $1\cdot41 \pm 0\cdot005$.

Tolerance

If you are making goods for a customer, the customer will often
dictate what leeway they will accept.

A customer may want bolts to fit nuts like the one shown here.

Ideally these bolts should have a diameter of 2 cm.

However, they will still do the job if they are a millimetre
out either way.

This leeway is called **tolerance**.

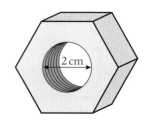

2 cm

In this example, the customer would ask for bolts of diameter $2 \text{ cm} \pm 0\cdot1 \text{ cm}$.

Example 1

A customer places an order for bolts of length $22 \text{ mm} \pm 0\cdot5 \text{ mm}$.

a **i** What is the maximum acceptable length of a bolt?
  ii What is the minimum acceptable length of a bolt?

b Which of these bolts is acceptable?
  i 22·3 mm   **ii** 21·4 mm   **iii** 22·6 mm

a **i** Maximum acceptable length of a bolt = 22 + 0·5 = 22·5 mm.
  ii Minimum acceptable length of a bolt = 22 − 0·5 = 21·5 mm.

b **i** Is acceptable.   **ii** Is too small.   **iii** Is too large.

Exercise 2.1A

1 Students were asked to measure an angle.
The marking instructions say, 'Accept the answer 21° ±2°.'

    a Which of these answers would be marked **incorrect**?

        20°   23°   19°   24°   21°   19°   21°   20°   18°   20°   21°   20°

    b What fraction of the answers were not within the acceptable tolerance?

2 The length of a wall is given as 5·5 metres, correct to 1 decimal place.

    a   **i** What is the shortest length that the wall could be?
        ii What is the longest length that the wall could be?

    b Its height is given as 1·2 metres, correct to 1 decimal place. State the range of possible heights.

3 *Top Notch* is a company that makes cashmere jumpers.
Jumpers marked 'Large' should have a chest measurement of 92 cm ±0·5 cm.

    a   **i** What is the maximum acceptable measurement?
        ii What is the minimum acceptable measurement?

    b Which of the jumpers in this batch would not pass quality control? They've been measured in centimetres to 1 decimal place.

        92·1   92·4   91·9   92·0   92·1   91·4   92·4   92·6   92·2   91·3

4 The weight of a packet of crisps is given as 25 grams to the nearest gram.
A sample of 10 packets of crisps is taken from a batch and weighed in grams correct to 1 decimal place.

    25·4   25·2   24·8   24·6   25·0   25·3   25·1   25·2   24·9   24·4

    a Which of the packets would be rejected?

    b What percentage of this sample is rejected?

Exercise 2.1B

1 A machine is set to make screws measuring 18 millimetres in length.
The acceptable margin of error is ±0·1 millimetres.
A sample of screws is taken from each batch and measured in millimetres to 2 decimal places.

    18·04   18·10   18·06   17·98   17·99   18·11   17·96   18·05   18·08   17·96

    a How many of the sample do not fall within the acceptable margin?

    b If the percentage of rejected screws is above 12%, the machine has to be reset.
       What advice would you give the company?

2 A piece of wood is measured, correct to 2 decimal places, as 3·46 metres.

    a Between what limits is the actual length of the wood?

    b A piece, measured as 1·27 metres, is cut off.
       What is the maximum possible length of this piece of wood?

    c What is the biggest possible length of the piece of wood that is left?
       (Take the smallest size from the longest possible original.)

    d What would be the minimum length of the piece left?

3 A box of matches says that the average content is 220 matches.

The company tests a sample of 10 boxes from every batch.

To satisfy quality control, the number of matches in each box must be 220 ±4, and the mean contents must be 220 (+ 0·5) matches per box.

Batch 1:   223   220   221   219   218   220   220   219   221   220

Batch 2:   220   221   220   218   221   219   217   216   222   220

Would both these batches pass quality control? If not, explain clearly why not.

4 Copper wire is manufactured for use in an electric circuit.

It should be 0·56 millimetres wide ±0·01 millimetres.

Twenty samples are taken and the standard of production checked.

The results of one sampling, measured in millimetres, are shown below.

What is the probability that a piece of copper wire, chosen from this sample at random, fails to reach the standard required?

0·555   0·562   0·566   0·571   0·560   0·565   0·563   0·559   0·549   0·555

0·558   0·570   0·564   0·569   0·563   0·557   0·556   0·565   0·567   0·555

2.2 Area

You should already know that:

$$\text{area of a triangle} = \tfrac{1}{2} \times \text{length of base} \times \text{altitude}.$$

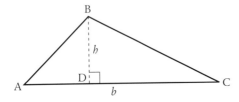

Shapes whose sides are straight lines can always be broken up into triangles when we want to find their area.

A farmer can calculate the area of his fields by dividing the fields into triangles and taking the right measurements.

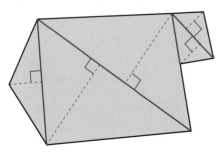

Other shapes may have special properties that will help you.

Example 1

Both the kite and the rhombus have an axis of symmetry.

Drawing this axis will split each shape into two congruent triangles.

Calculate the area of both shapes.

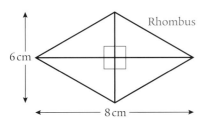
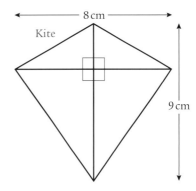

The diagonals of both the kite and rhombus intersect at right angles.
One diagonal lies on an axis of symmetry.

Drawing the axis of symmetry bisects the other diagonal ...

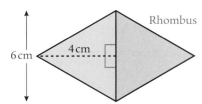
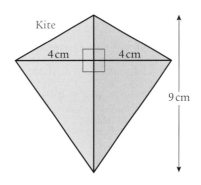

Both triangles have a base of 6 cm and an altitude of 4 cm.

Area of green triangle $= \frac{1}{2} \times 6 \times 4$
$= 12 \text{ cm}^2$.

Area of purple triangle = area of green triangle.

So area of rhombus $= 2 \times 12 = 24 \text{ cm}^2$.

Both triangles have a base of 9 cm and an altitude of 4 cm.

Area of green triangle $= \frac{1}{2} \times 9 \times 4$
$= 18 \text{ cm}^2$.

Area of purple triangle = area of green triangle.

So area of kite $= 2 \times 18 = 36 \text{ cm}^2$.

Example 2

A parallelogram has half-turn symmetry.

Calculate the area of this parallelogram.

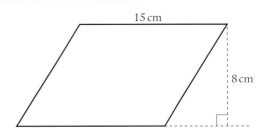

Drawing a diagonal will split the parallelogram into two congruent triangles.

The base of a triangle will be a side of the parallelogram.

The altitude will be the perpendicular distance to the opposite side.

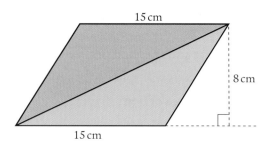

Area of green triangle $= \frac{1}{2} \times 15 \times 8 = 60 \text{ cm}^2$.

Area of purple triangle = area of green triangle.

So area of parallelogram $= 2 \times 60 = 120 \text{ cm}^2$.

Example 3

A trapezium has a pair of parallel sides.

A line perpendicular to one is perpendicular to the other.

Calculate the area of the trapezium.

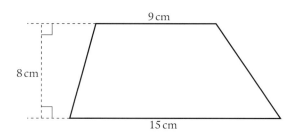

Drawing a diagonal we get two triangles:

... a green one with a base of 9 cm and an altitude of 8 cm.

$$\text{Area of green triangle} = \frac{1}{2} \times 8 \times 9$$
$$= 36 \text{ cm}^2.$$

... a purple one with a base of 15 cm and an altitude of 8 cm.

$$\text{Area of purple triangle} = \frac{1}{2} \times 8 \times 15$$
$$= 60 \text{ cm}^2.$$

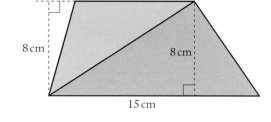

Area of the trapezium is the sum of the areas of the green and purple triangles $= 36 + 60 = 96 \text{ cm}^2$.

Exercise 2.2A

1 Calculate the area of each of these quadrilaterals.

a

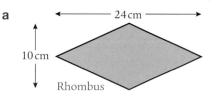

Rhombus

24 cm
10 cm

b

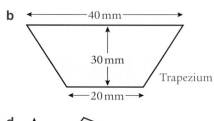

Trapezium

40 mm
30 mm
20 mm

c Parallelogram

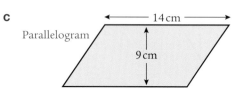

14 cm
9 cm

d

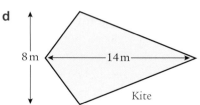

Kite

8 m
14 m

2 Suki has a metal construction set.

She uses two 12-centimetre and two 8-centimetre lengths to make a rectangle.

However, the shape moved and the shape became a parallelogram, with height 7 centimetres.

Which of the two shapes has the larger area and by how much?

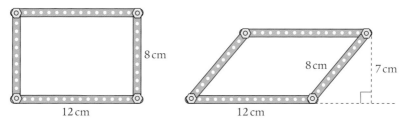

3 A section of a bridge is made from girders in the shape of a rhombus.

Calculate the area within one section of the bridge.

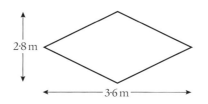

4 A house and garden are situated on a hill.

A retaining wall is needed to stop the soil from the garden falling on to the pavement.

The wall is trapezium shaped.

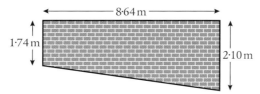

a Calculate the area of the face of the wall.

b The bricks will cost £30 per square metre. Mortar, concrete and top stones will cost £335.
Labour for the job has been priced at £690.
Calculate the total cost of building the retaining wall.

5 A section of floor is decorated by congruent tiles in the shape of a kite.

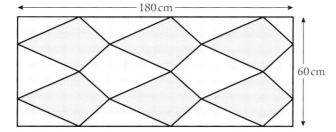

a Find the area of one tile by first finding its diagonals.

b **i** If the rectangular area were to be cut up and reformed into kites, how many kites would you get?

   ii Find the area of the rectangle and hence the area of one kite.

c Which strategy is easier for finding the area of one kite? Explain why.

Formulae for finding area

Using the above methods for finding the areas of quadrilaterals, formulae can be formed.

Rhombus and kite

Both shapes have diagonals of length d_1 and d_2 units:

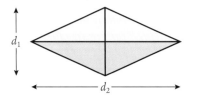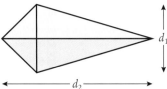

Area of yellow triangle $= \frac{1}{2} \times$ base $\times$ height $= \frac{1}{2} \times d_2 \times \frac{1}{2} \times d_1 = \frac{1}{4} d_1 d_2$.

Area of rhombus or kite $= 2 \times$ area of yellow triangle $= \frac{1}{2} d_1 d_2$.

Area of rhombus or kite = half the product of the diagonals.

Parallelogram

Let the parallelogram have a base of a units and an altitude of b units.

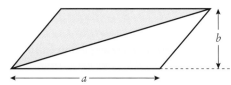

Area of parallelogram $= 2 \times$ area of yellow triangle $= 2 \times \frac{1}{2} \times a \times b = ab$.

Area of parallelogram = base × altitude.

Trapezium

Let the trapezium have parallel sides of a units and b units, which are h units apart.

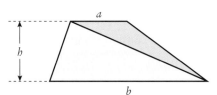

Area of trapezium $=$ area of green triangle $+$ area of yellow triangle $= \frac{1}{2} ah + \frac{1}{2} bh = \frac{1}{2} h (a + b)$.

Area of trapezium = half sum of parallel sides × distance between them.

Exercise 2.2B

1   Sketch each shape described and use the appropriate formula to find the area of the shape.

  a   A parallelogram with a base of 16 cm and an altitude of 9 cm.

  b   A trapezium with parallel sides of length 12 cm and 16 cm, which are 9 cm apart.

  c   A kite with diagonals of length 28 mm and 47 mm. (Does it matter where they cross?)

  d   A rhombus with diagonals 22 cm and 18 cm.

  e   A rhombus with a base of 10 cm and an altitude of 5 cm. (Every rhombus is a parallelogram.)

2 A 'Grow Your Own' herb kit is packaged in boxes with a front face in the shape of a trapezium.

Calculate the area of the front face of the box.

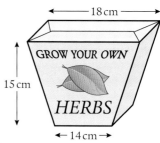

3 A toy making company makes kites of different sizes.

The diagonals of the *Osprey* measure 50 centimetres by 24 centimetres.

The diagonals of the *Merlin* measure 44 centimetres by 28 centimetres.

One factor that influences a kite's ability to fly is its area.

Which of these two kites has the bigger area?

4 A rectangle has its dimensions given as 3·0 cm and 4·0 cm measured to 1 decimal place.

Remember this means the dimensions are 3·0 ±0·05 cm and 4·0 ±0·05 cm.

a Between which two values does the area of the rectangle lie?

b A kite has diagonals 3 cm and 4 cm, measured to the nearest centimetre.

Between which two values does the area of the kite lie?

5 A rectangular sheet of metal measures 5 metres by 0·5 metre.

Rhombus-shaped plates have to be cut from the sheet of metal.

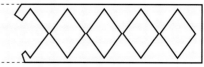

The diagonals of each rhombus measure 34 centimetres and 48 centimetres.

a How many rhombuses can be cut from a single sheet of metal?

b Calculate the area of one rhombus.

c What area of metal is wasted?

d If the amount of wastage is above 52%, the company would look to changing the size of the sheet of metal they use.

Should the company be looking to source sheets of metal of a different size?

Justify your answer.

6 An alphabet book for small children is designed so that it folds in a zigzag way.

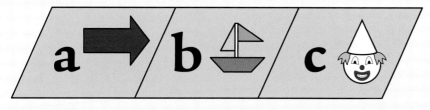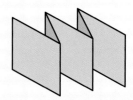

Each page is in the shape of a parallelogram measuring 12 centimetres long by 10 centimetres high.

A letter and a picture are printed on each side of a page.

a What area of cardboard is needed to make the alphabet book?

   b If, instead, they had made the book from a 12-centimetre by 10-centimetre rectangle, what area of cardboard would have been needed?

   c Give a reason why they might choose to manufacture:
      i the rectangular design
      ii the parallelogram design.

   (Hint: try making a smaller version of each book from scrap paper.)

7 Aleta and Eddie move into a new house and discover their lawn needs to be replaced.

They want to re-lay the lawn and place edging stones round it.

The cost of turf, preparing and laying the lawn is quoted as £5·25 per square metre.
Edging costs 54 pence per metre.

The lawn is a trapezium with dimensions as shown.

   a Calculate how much it will cost them to re-lay and edge their lawn.

   b Their next-door neighbour, whose lawn is in the shape of a parallelogram, discovers it has the same area as that in Aleta and Eddie's garden.

   The front edge of their lawn is 16 metres.
   How far is it to the back edge of the lawn?

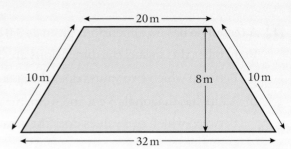

2.3 Circles: a class activity

1 Collect several circular objects.

2 Measure both the diameter and the circumference of each object,
recording your findings in a table. (You could add to the table below.)

Object	Diameter, D (cm)	Circumference, C (cm)
Roll of tape	25·4	80·0
Butter tub	10·4	32·5
Lid	3·7	11·6
Roll of tape	9·1	28·6
Biscuit	4·3	13·6
Small dish	12·1	38
£2 coin	2·8	8·9
£1 coin	2·4	7·5
Cup	7·5	23·5
Tin can	7·8	24·5

3 Draw a graph of **diameter** against **circumference**.
You should find all points lie on a straight line, passing through the origin.
When this happens in maths it means that one measurement is just a multiple of the other.

4 Find the factor. In this case you can see that the circumference (C cm) is roughly three times the diameter (D cm).

More accurately, we find that $C \approx 3.14D$.

To be exact, we use the symbol π, and write $C = \pi D$.

The symbol π represents the actual number of times that the circumference is bigger than the diameter ... it turns out to be 3·1415926535897932384 ..., a decimal which never ends or recurs.

Generally we use $\pi = 3.14$ (to 2 d.p.) for most of our calculations

... or $\pi = \frac{22}{7} = 22 \div 7$, if we don't have a calculator

... or the **correct** button if we do have a calculator ...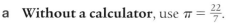

$C = \pi D$

Example 1

A can of *Sunshine Sweetcorn* has a diameter of 63 millimetres.

Calculate the circumference of the can:

a without the aid of a calculator

b with a calculator.

Give each answer correct to 1 decimal place.

a **Without a calculator**, use $\pi = \frac{22}{7}$.

$C = \pi D$.

$\Rightarrow \quad C = 63 \times \frac{22}{7} = 63 \div 7 \times 22 = 9 \times 22 = 198$ cm.

b **With a calculator**, use the π button.

$C = \pi D$.

$\Rightarrow \quad C = 63 \times \pi = 197 \cdot 920337... = 198 \cdot 0$ mm (correct to 1 decimal place).

Exercise 2.3A

1 Without the aid of a calculator, find the circumference of a circle:

   a with **diameter** i 49 cm ii 2·1 m iii 350 mm

   b with **radius**   i 14 cm ii 5·6 m iii 7 mm.

2 Calculate the circumference of each of these circular objects.

   You can use the π-button on your calculator or let $\pi = 3.14$.

a

b

c
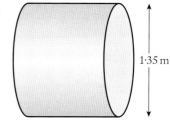

3 The centre part of a stained glass window, in a church, is in the shape of a circle.

The radius of the circle is 36 centimetres.

The lead that keeps the circle in place needs to be replaced.

What length of lead is required?

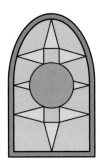

4 A lampshade has a circular top and base, with radius 9·6 centimetres and 19·7 centimetres respectively.

Kirsty decides to put green trim round both the top and base of the lampshade, so that it will match her new curtains.

What length of trim will she need to complete the job?

5 Rachel is making a sponge birthday cake.

The recipe says to use a 7-inch round cake tin (i.e. the diameter is 7 inches).

Rachel has bought the ribbon to go round this size of cake.

However, she doesn't have a tin of that size.

She does have a 6-inch square tin, a 6-inch round tin and a 6-inch by 7-inch rectangular tin.
Explain which tin she should use.

6 If $C = \pi$ times D, then $D = C \div \pi$. $\left(D = \dfrac{C}{\pi}\right)$

   a Calculate, with the aid of a calculator, the diameter of a circle of circumference:
     i 20 cm
     ii 234 mm.

   b To divide by π (approximately) we can divide by 22 and multiply by 7.

     Without a calculator, find the diameter of a circle of circumference:
     i 44 cm
     ii 8·8 cm
     iii 11 cm.

Exercise 2.3B

1 The circumference of a coffee table is 2·26 metres.

By working backwards, calculate the diameter of the coffee table.

2 A trundle wheel is used to measure long distances where it would be impractical to use a tape.
Each rotation of the wheel measures 1 metre along the ground.

Calculate the radius of the wheel, to three significant figures.

3 A Penny Farthing bicycle has wheels of different sizes.
The front wheel has a radius of 30 inches and the back wheel a radius of 9 inches.

    a Calculate the circumference of the front wheel.

    b How many times will the front wheel have to turn to travel one mile? (1 mile = 63 360 inches.)

    c Express the number of times the back wheel turns in a mile as a multiple of the number of times the front wheel turns in a mile.

4 Big Ben is the nickname for the great bell that hangs in the Elizabeth Tower at Westminster.
When Big Ben was cast, it was the biggest bell in the world.

It is 2·29 metres tall and the circumference of the base of the bell is 8·61 metres.

    a Calculate the diameter of the base of Big Ben.

    b The hands on each clock on the Elizabeth Tower measure 1·9 metres and 3·5 metres from the centre of the clock face to the tip of the hand.

      i How far will the tip of the minute hand travel in one hour?

     ii How far will the tip of the hour hand travel in one hour?

    iii How many times faster is the tip of the minute hand travelling compared to the tip of the hour hand?

2.4 Composite shapes – perimeters

Many shapes have perimeters which are a mix of straight lines and parts of a circle.

We find the perimeter by carefully considering each part of the outline.

Example 1

A stadium has a perimeter in the shape of a rectangle with a semi-circle at one end.

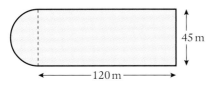

The straights are 120 m long, separated by 45 m.

What is the perimeter of the stadium?

The stadium is made of two straights of length 120 m and one of 45 m plus a semicircle of diameter 45 m.

Perimeter $= 2 \times 120 + 45 + \frac{1}{2} \times \pi \times 45$

$\qquad\qquad = 285 + 70{\cdot}685...$

$\qquad\qquad = 355{\cdot}7$ m (to 1 d.p.).

The perimeter of the stadium is 355·7 m (to 1 d.p.).

Exercise 2.4A

Round your answer to 3 significant figures unless told otherwise.

1 Calculate the perimeter of each of these composite shapes.

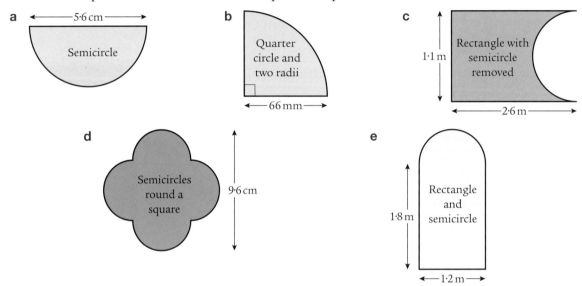

a ←—— 5·6 cm ——→ Semicircle

b Quarter circle and two radii ←— 66 mm —→

c 1·1 m | Rectangle with semicircle removed ←— 2·6 m —→

d Semicircles round a square | 9·6 cm

e 1·8 m | Rectangle and semicircle ←— 1·2 m —→

2 The top of a breakfast bar is a rectangle with a semicircle on the end.
It is placed against a wall.

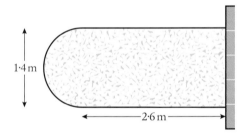

1·4 m

←—— 2·6 m ——→

It is suggested that to sit comfortably round a table, each person needs 65 centimetres of space.
Can 10 people sit comfortably round this breakfast bar?
Explain your decision.

Exercise 2.4B

1 A garden pond is in the shape of a rectangle with semicircular ends.

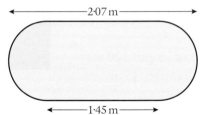

←———— 2·07 m ————→

←— 1·45 m —→

a Work out the radius of each of the semicircular ends.
b Calculate the perimeter of the pond, to 2 decimal places.

c The owners want to put a strip of coloured lights round the pond.

They are advised to buy a length 5% longer than the perimeter of the pond.

The lighting costs £12·30 per metre.

How much will the lighting cost altogether?

2 Draught excluder tape is fitted round the edge of a door to stop the draught getting in.

The door is a composite shape, made up of a rectangle and a semicircle.

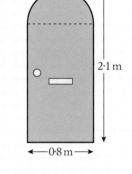

a What is the diameter of the semicircle?

b Calculate the height of the rectangular part of the door.   (Hint: it is not 2·1 metres.)

c Calculate the length of tape required for this door.

2·1 m

0·8 m

3 The perimeter of this shape is 256 cm.

The shape is made up of a rectangle and a semicircle.

The width of the shape is 36 cm.

Calculate the total length of the shape.

36 cm

2.5 Area of a circle

A circle, with radius r, has been cut into 16 congruent sectors.

The sectors have been rearranged to make a shape that is very similar to a parallelogram.

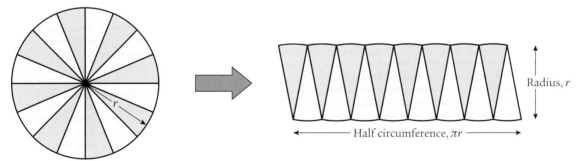

Radius, r

Half circumference, πr

If the circle were cut into even thinner sectors, the rearranged shape would get even closer to a parallelogram.

The parallelogram that is produced has a base equal to half the circumference of the circle,

i.e. $\pi \times \dfrac{d}{2} = \pi \times r$. The altitude of the parallelogram is the radius of the circle, r.

So the area of the circle becomes the same as the area of the parallelogram:

$A = \text{base} \times \text{altitude} = \pi r \times r = \pi r^2$.

Area of a circle, $A = \pi r^2$.

Example 1

The Floral Clock in Princes Street, Edinburgh, has a radius of 3·4 m.

The inner ring has a diameter of 3·4 m.

Calculate the area of:

a   the clock face   b   the inner ring.

a   Area of clock face, $A = \pi r^2 = \pi \times 3\cdot4^2 = 36\cdot316811... = 36\cdot3 \text{ m}^2$ (to 1 d.p.).

b   If the diameter of the inner ring is 3·4 m then the radius is 1·7 m.   (3·4 ÷ 2)
   $A = \pi r^2 = \pi \times 1\cdot7^2 = 9\cdot079202... = 9\cdot1 \text{ m}^2$ (to 1 d.p.).

Exercise 2.5A

1   Calculate the area of a circle with:

   a   radius 14 cm   b   radius 3·45 m   c   radius 98 mm

   d   diameter 144 mm   e   diameter 23·7 cm.

2   Calculate the area of each circular shape.

a

b
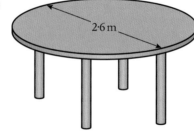

c

3   A replacement piece of glass is need for a corner shelving unit.

The shelf is a quarter circle with radius 17·4 centimetres.

Calculate the area of the replacement glass.

4   A semicircular shaped flower bed with diameter 4·2 metres requires fertiliser to help the plants grow.

A bottle of fertiliser costs £3 and covers 1·5 m².

How much will it cost to treat the flower bed?

Example 2

The end of a tunnel in a model railway is a rectangle with a semicircle removed.

Calculate the yellow area.

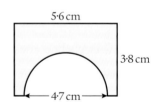

Area = rectangle − semicircle
$$= 5\!\cdot\!6 \times 3\!\cdot\!8 - \frac{1}{2} \times \pi \times 2\!\cdot\!35^2$$
$$= 21\!\cdot\!28 - 8\!\cdot\!67472...$$
$$= 12\!\cdot\!6 \text{ cm}^2 \text{ (to 1 d.p.)}.$$

The yellow area is $12\!\cdot\!6$ cm² (to 1 d.p.).

Exercise 2.5B

1 The pressure on a washer in a tap is proportional to its area.

Calculate the area of each washer.

a

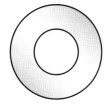

external radius: 5 cm
internal radius: 2 cm

b

external diameter: 2 cm
internal diameter: $1\!\cdot\!2$ cm

2 Calculate the coloured area in each shape:

a

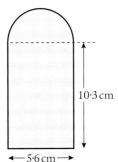

10·3 cm

←5·6 cm→

Rectangle and semicircle

b

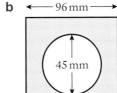

← 96 mm →

45 mm

Square and circle

c

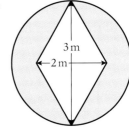

3 m

←2 m→

Rhombus and circle

3 *Fun Felt Creations* is a company that makes brooches and decorations from felt.
One of their brooch designs consists of two layers of felt held together by a button.
Each layer is in the shape of a square with sides 32 mm long surrounded by four semicircles, as shown in the diagram.
Calculate the area of felt used to make one brooch.

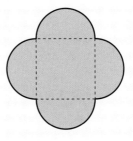
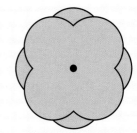

4 This sign indicates that bicycles are not to be ridden.

The diameter of the sign is 750 millimetres and the red ring is 75 millimetres wide.

a Estimate what percentage the area of the red ring is of the whole shape.

b Calculate the area of the sign.

c Calculate the area of the red ring.

d Work out what percentage the area of the red ring is of the whole shape.
Comment on your answer.

2.6 Prisms

A prism is a solid with a uniform cross-section.

Hexagonal prism

Circular prism = cylinder

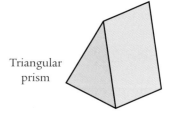

Triangular prism

It takes its name from the shape of its base.

It can be shown that:

volume of prism = area of base × distance between ends.

Example 1

A confectioner wishes to make a box for his sweets.

Which of these boxes has the bigger volume?

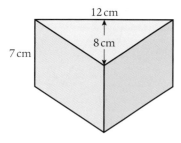

12 cm

8 cm

7 cm

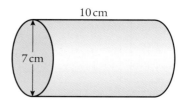

10 cm

7 cm

Volume of triangular prism

    = area of base × distance between ends

    = $\frac{1}{2}$ × 12 × 8 × 7

    = 336 cm³

Volume of cylinder

    = area of base × distance between ends

    = π × 3.5^2 × 10

    = 384·8 cm³ (to 1 d.p.)

So the cylinder has the bigger volume.

Exercise 2.6A

1 Calculate the volume of each of the prisms.

a
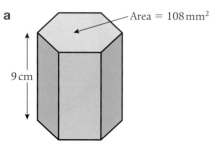
Area = 108 mm²

9 cm

b
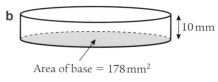
10 mm

Area of base = 178 mm²

c
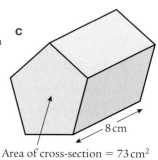
8 cm

Area of cross-section = 73 cm²

2 Calculate the volume of each prism by first finding the area of the base.

a

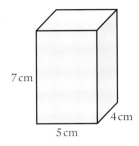

7 cm
5 cm
4 cm

b

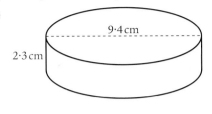

9·4 cm
2·3 cm

c

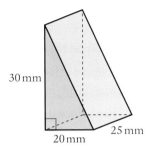

30 mm
20 mm
25 mm

3 A box for chocolates has a trapezium cross-section and a length of 19 centimetres.

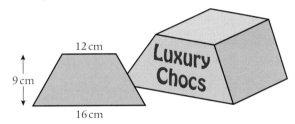

12 cm
9 cm
16 cm

The dimensions of the trapezium are shown in the diagram.

Calculate the volume of the box.

4 *Wax Ways* is a company that makes scented candles.

The candles in one set are cylinders in three different sizes.

The tall candle: 22·4 centimetres tall with a diameter of 6·8 centimetres.

The wide candle: 9·8 centimetres tall with a diameter of 12·4 centimetres.

The small candle: 9·8 centimetres tall with a diameter of 6·8 centimetres.

a Calculate the volume of the tall candle, to the nearest cubic centimetre.

b What is the difference in volume between the wide candle and the small candle?

c The cost of the candle depends on its volume.

   The tall candle costs £12·20. How much does the small candle cost?

d A cubic centimetre of wax weighs 0·9 grams.

   What weight of wax is needed to make one set of candles?

5 Ornaments are packed into 5-centimetre cube boxes.

These boxes are packed into a larger carton, which measures 20 centimetres by 20 centimetres by 15 centimetres.

How many boxes can be fitted in one carton?

6 *Sunshine Cereal* boxes are packed into large boxes for transportation to the shops.

The boxes of cereal measure 30 centimetres by 20 centimetres by 10 centimetres.

How many packs of cereal will fit in the box?

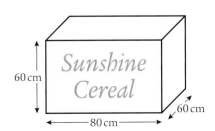

Sunshine Cereal
60 cm
80 cm
60 cm

Exercise 2.6B

1 A metal water trough has a trapezium-shaped cross-section.
It is 1·8 metres long.

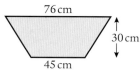
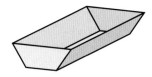

a Calculate the volume of the trough when full.

b Find the capacity of the trough in litres.
($1000 \, cm^3 = 1$ litre.)

c It is estimated that a sheep will drink 4 litres of water per day.
The trough is in a field that contains 27 sheep.
Jill fills the trough on Monday morning. Would she be able to next fill it on Thursday morning without endangering the sheep?
Explain your answer thoroughly.

2 £1 coins are made of a nickel–brass alloy.
The £1 coin has a diameter of 22·5 millimetres and has a thickness of 3·15 millimetres.

a Calculate the volume of a £1 coin.

b How many £1 coins could be minted from a cubic block of alloy with sides 200 millimetres?

c What is the total weight of these coins if one £1 coin weighs 9·5 grams?

3 Asha is making individual vegetable pies for a party. The pies have a pastry top.
Puff pastry comes in a block that measures 16 centimetres by 10 centimetres by 2·2 centimetres.
It is rolled out to a thickness of 0·8 centimetre.
Circles of diameter 5·8 centimetres are then cut from it.
Any pastry left is gathered up and rolled out again until no more complete circles can be cut.
How many pies can Asha make from this block of pastry?

4 These two prisms have the same volume.

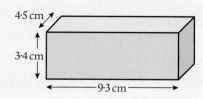
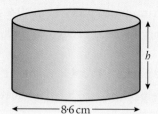

Calculate the height of the cylinder, correct to 1 decimal place.

5 Hollow steel tubing has an outer diameter of 89 millimetres.
The metal is 3 millimetres thick.
Calculate the volume of metal in a two-metre length post made from this tubing.

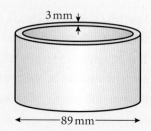

6 In a factory, cans of soup are packed into boxes like these to be sent to the distributers.

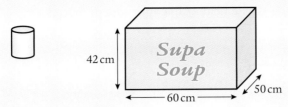

42 cm

Supa Soup

60 cm · 50 cm

The cans of soup measure 7 centimetres in diameter and are 10·4 centimetres high.

  a What is the volume of a can?

  b How many cans can fit in the box?

  c What is the volume of empty space in the box?

Preparation for assessment

1 Calculate the area of each shape.

  a   38 mm   83 mm   **b** 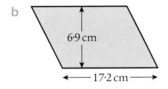 6·9 cm   17·2 cm   **c** 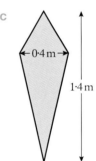 0·4 m   1·4 m

2 A circle has a radius of 4·6 centimetres.

  a Calculate its circumference.

  b Find the area of the circle.

3 A trapezium has parallel sides measuring 14 centimetres and 19 centimetres.

  The distance between the parallel sides is 9·5 centimetres.

  Calculate the area of the trapezium.

4 For each solid (all measurements are in centimetres):

  i draw its net

  ii write in the sizes of the sides

  iii work out the surface area

  iv calculate the volume.

  a 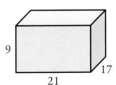 9   17   21   **b** 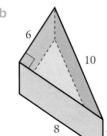 6   10   8   2

5 A farmer has an L-shaped field surveyed.

   a Calculate the perimeter of the field. (All lengths are in metres.)

   b Calculate its area.

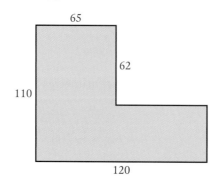

6 The water sprinkler rotates, watering a region of the lawn that is three-quarters of a circle.

   The radius of the circle is 25 metres.

   The lawn is a rectangle, 60 metres by 80 metres.

   Calculate:

   a the perimeter of the shape watered

   b the area of the shape watered

   c the percentage of the lawn watered.

7 A rubber washer for a tap is 18 millimetres wide.

   The diameter of the hole in the middle is 5 millimetres.

   a Calculate the area of the circular face of the washer.

   b The washer is 7 millimetres thick.

     Calculate the volume of the washer.

8 A toy clown's face is made from a semicircle of wood with an isosceles triangle for a hat.

   The dimensions, in centimetres, are shown in the diagram.

   a Calculate the area of the shape.

   b The wood is 1·2 centimetres thick.

     Calculate the volume of the toy clown face and hat.

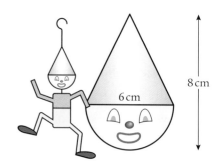

9 Cylindrical cans with diameter 7·8 centimetres and height 12·2 centimetres are fitted into a box.

   The box measures 90 centimetres by 90 centimetres by 40 centimetres.

   a Work out how many cans can fit into the box.

   b Calculate the volume of one can.

   c Calculate the volume of the box.

   d What percentage of the box is empty?

10 Remember Rachel's plans to make a horse riding arena?

She has 200 metres of temporary fencing to make the arena.

She wants the arena to have the largest possible area.

What shape would be best for the arena?

What would be the dimensions of the arena?

3 Gradients and straight lines

⏸ Before we start...

You are going on holiday to Australia and need to change some money.

Different companies charge different amounts for the service.

Holiday Cash
RATE: £1 BUYS AUS $1·60
COMMISSION: 10%

Xchange
RATE: £1 BUYS AUS $1·50
COMMISSION: £5 PER TRANSACTION

COMMISSION
CHARGED BEFORE
EXCHANGE OCCURS

Which is the best company for you?

Investigate the situation.

▶ What you need to know

1 Express each fraction in its simplest form.

   a $\frac{15}{18}$ b $\frac{60}{140}$ c $\frac{84}{300}$

2 Write each common fraction: **i** in decimal form **ii** in ratio form **iii** as a percentage.

   a $\frac{7}{10}$ b $\frac{2}{5}$ c $\frac{3}{8}$

3   In an aquarium there are 10 goldfish and 18 tropical fish.

Write down in its simplest form, the ratio of:

a goldfish to tropical fish

b goldfish to fish.

4   **a** State the coordinates of A and B.

b How much to the right of A is B?

c How much above A is B?

d What do the arrowheads in the diagram indicate?

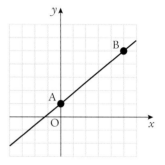

3.1 Slope and gradient

Calculating the gradient

In mathematics the **gradient** is a measure of the 'steepness' of a slope.

It is defined as the ratio of the vertical change to the horizontal change as you move between two points on the slope.

The gradient between the points A and B is defined as:

$$\text{gradient}_{AB} = \frac{\text{vertical change}}{\text{horizontal change}}.$$

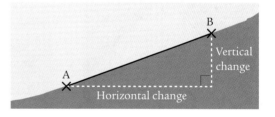

On the Forth Bridge, the rail track is **horizontal** and the main supports are **vertical**.

We define the gradient of **horizontal** lines as zero.

We say the gradient of **vertical** lines is undefined.

Example 1

Mount Teide is a volcano on Tenerife.

A vulcanologist surveying the slope, measures the distance horizontally and vertically between two points. The data is shown in the diagram. Calculate the gradient of the slope.

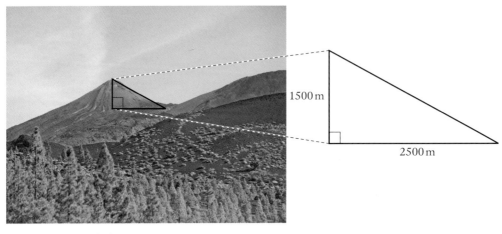

$$\text{Gradient} = \frac{\text{vertical change}}{\text{horizontal change}} = \frac{1500}{2500} = \frac{3}{5} = 0.6.$$

The gradient of this slope can be quoted as 0·6, or 3 in 5, or 60%.

Exercise 3.1A

1 Calculate the gradient for each roof, leaving your answer as a fraction and simplifying where appropriate.

a

b

c

2 a Calculate the gradient of each ladder as a decimal fraction.

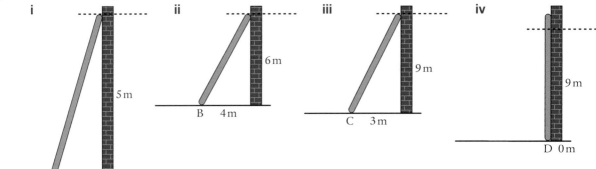

b Health and Safety regulations state:

'For every 4 units up the wall, place the ladder foot 1 unit out from the wall.'

Any steeper would be considered dangerous.

Which ladder positions in part **a** are not safe?

3 A rollercoaster is a rail supported by struts.

Calculate the gradient of the struts (drawn in red) holding the different sections of the rollercoaster:

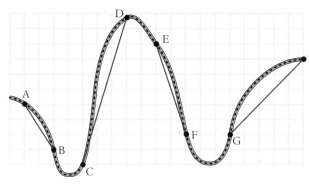

| **a** AB | **b** CD | **c** EF | **d** GH. |

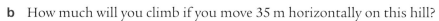

Example 2

Road signs are used to warn motorists of steep hills.

They give the gradient as a percentage or a ratio.

A road sign at the start of a hill indicates its steepness using the ratio $1:7$.

a Express this as:

   i a common fraction   **ii** a percentage.

b How much will you climb if you move 35 m horizontally on this hill?

c Is a $1:5$ hill steeper or shallower than a $1:7$ hill?

a **i** $1:7 = \frac{1}{7}$.

   ii $\frac{1}{7} = \left(\frac{1}{7} \times 100\right)\% = 14\%$.   To the nearest whole number … road signs don't do decimals

b $1:7$ is a rise of 1 unit for every 7 units forward.

   If you go 35 m forward, then you climb $\frac{35}{7} = 5$ m up.

c Express $1:5$ as a percentage to make comparison easier:

   $\frac{1}{5} = \left(\frac{1}{5} \times 100\right)\% = 20\%$.

   $20\% > 14\%$

   So a $1:5$ hill is steeper than a $1:7$ hill.

Exercise 3.1B

1 a Which of the following roads is the steepest?
  i Largiebeg to Largiemhor ... a climb of 0·15 km over 33 km.
  ii Ardtor to Balmorin ... a climb of 0·2 mile over 5 miles.
  iii Corrie to Dunelk ... a climb of 250 m over 3·8 km.

b Express each of the gradients in part **a** as a percentage.

c A hill will only merit a road sign if it has a slope of 5% or more.
  Which of the roads will have a road sign?

2 A stretch of road near Applecross has a gradient of 1 in 5.

a Rewrite this gradient as: **i** a fraction and **ii** a percentage.

b A hill near Loch Ness has a gradient of 15%.
  How does this hill compare with the one near Applecross?

c How much height would a car gain if its horizontal change was 300 m as it drove up the stretch of road near Applecross?

3 The Health and Safety guidelines for wheelchair ramps in buildings depend on the horizontal distance.

For a dwelling (place where people live) the guidelines are as follows.

- Horizontal distance less than 10 m can have a gradient of 1 : 15 or less.

- Horizontal distance 5 m or less can have a gradient of 1 : 12 or less.

a From the measurements given, do these ramps fit with the guidelines?

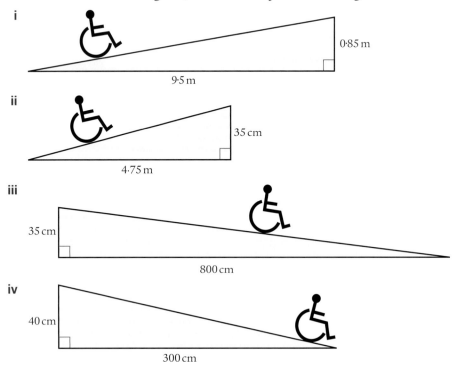

i

9·5 m   0·85 m

ii

4·75 m   35 cm

iii

35 cm   800 cm

iv

40 cm   300 cm

b If the horizontal distances are fixed for each of the ramps above, what changes need to be made to the vertical distance for them to fit with the guidelines?

4 Ski slopes in the USA are classified according to their gradient.

Each class is given a name that is easy to signpost.

Slope name	Gradient range
Green circle	6% to 25%
Blue square	25% to 40%
Black diamond	> 40%

a What is the gradient of each of the slopes in the table?
   i a run that drops 0·85 km over 5 km.
   ii a run that drops 1·2 km over 3·3 km.

b In what class would each slope be placed in the USA?

c A skier on slope **ii** travelled a horizontal distance of 1·5 km.
   How far did she descend in this time?

d A skier descended 15 m on slope **i**.
   What horizontal distance did he move?

   (Round your answers to **c** and **d** to 2 decimal places.)

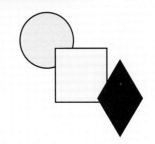

5 The Ordnance Survey key to steepness uses arrowheads placed along the road.

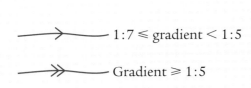

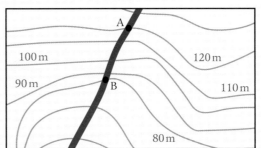

The map shows a 200 m stretch of road, AB, crossing the contours.

We can see from the contours that between A and B the road drops 40 m.

Gradient of this road $= \frac{40}{200} = \frac{1}{5} = 0·2$ ... a gradient of 1 in 5.

This road should have a double arrowhead road marking.

What symbol, if any, would be used for the following stretches of road?

a PQ = 250 m   **b** RS = 130 m   **c** TU = 600 m

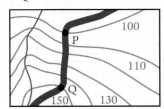   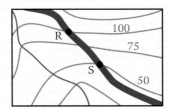   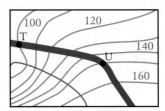

6 Find out about local building plans in your area and investigate the safety regulations about gradients of different things about the house, e.g. stairs, roofs, ramps, paths, drain pipes.

Do the same rules apply for public buildings?

Go on the internet and research the safe use of ladders and ramps.

Local planners have decision rules in place as to when a hill merits a road sign.
It's not just as simple as 'over 5%'.

Find out more.

3.2 Graphs and gradients

In coordinate diagrams, we don't use the terms 'vertical' and 'horizontal'.

Instead we talk of the y-direction and the x-direction.

We still discuss gradient, however.

By tradition the letter m has been used to represent the gradient of a line.

The gradient between two points A and B is represented by m_{AB}.

The definition of the gradient is redefined as:

$$m_{AB} = \frac{\text{change in } y\text{-direction}}{\text{change in } x\text{-direction}}$$ as you move in the x-direction.

For example, the gradient from A to B in the diagram is:

$$m_{AB} = \frac{4}{6} = \frac{2}{3} = 0.66\dot{6} \text{ or } 66\frac{2}{3}\%.$$

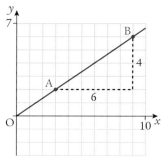

Line graphs are often used to highlight the relationship between two variables.

The gradient of the line is a measure of how the y-measurement changes as the x-measurement changes ... the rate of change of y per unit x.

For example, if we plotted a graph of the cost in pounds of a consignment of oranges (y-axis) against the number of oranges bought (x-axis), then the gradient would give us the number of pounds per orange ... the rate being charged.

If we plotted distance travelled in kilometres against time in hours, the gradient would give us the number of kilometres travelled per hour ... the speed.

Example 1

The graph below can be used for converting pounds sterling (British pounds) to and from euros.

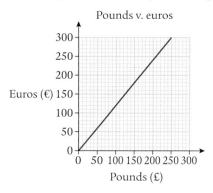

a How many euros would you get for £250?

b How many pounds sterling would you get for €150?

c By calculating the gradient of the line, find the exchange rate for changing £s to €s.

René Descartes, a Frenchman, first came up with the coordinate system and it is believed that he used the letter *m* for the gradient, taking it from the French word *monter*, which means 'to go up' or 'climb'.

a From the graph, you would get €300 for £250.

b From the graph, you would get £125 for €150.

c Pick any two convenient points on the line ... use the points from parts **a** and **b**.

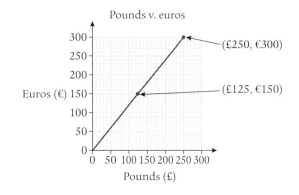

Pounds v. euros

(£250, €300)

(£125, €150)

Euros (€)

Pounds (£)

$$m = \frac{\text{change in } y\text{-direction}}{\text{change in } x\text{-direction}}$$

$$= \frac{€(300 - 150)}{£(250 - 125)}$$

$$= \frac{€150}{£125} = \frac{€1 \cdot 2}{£1}.$$

So the exchange rate is €1·20 per £1.

Galileo Galilei studied the motion of falling bodies.

He knew that, from a graph of **distance** against **time**, he could calculate the **speed** of any section of a journey by finding the gradient.

Example 2

The graph illustrates a journey to the shops and back home again by a family.

a How long did it take to get to the shops?

b How far away are the shops from home?

c How long was spent at the shops?

d How can you tell which leg of the journey was fastest?

e Calculate the speed they travelled at when going to the shops.

f Calculate the speed they travelled at when coming home from the shops.

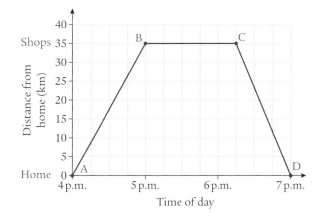

a It took 1 hour to get to the shops.   From 4 p.m. to 5 p.m. is one hour

b The shops are 35 km from home.

c The horizontal line represents the time spent at the shops.   Gradient = 0 ⇒ speed = 0
From 5 p.m. to 6.15 p.m. is 1 hour and 15 minutes or 1·25 hours.

d The line representing the journey home is steeper, so it must be the faster journey.

e To calculate the speed, we need to calculate the gradient from point A to point B:

$$m_{AB} = \frac{\text{change in } y}{\text{change in } x} = \frac{35 \text{ km}}{1 \text{ h}} = 35 \text{ km/h}.$$

f To calculate the speed, we need to calculate the gradient from point C to point D:

$$m_{CD} = \frac{\text{change in } y}{\text{change in } x} = \frac{35 \text{ km}}{0 \cdot 75 \text{ h}} = 46 \cdot 667 \text{ km/h}.$$

49

Exercise 3.2A

1 The distance–time graph represents the Number X95 bus travelling from the depot.

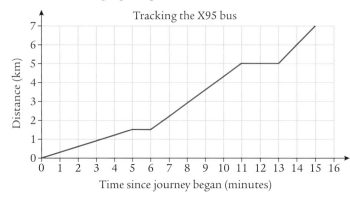

Tracking the X95 bus

Distance (km) / Time since journey began (minutes)

Stance 9

Hawick X95

a How far is it to the first bus stop?

b For how long does the bus wait at the second stop?

c During which section of the trip was the bus travelling fastest?

d Calculate the speed at which the bus was travelling after the second stop:
   i in km per minute   **ii** in km per hour.

2 Petrol in the UK is now sold in litres. It was once sold in gallons.
We can convert between gallons and litres using the fact that,
to 2 decimal places, 1 gallon is 4·55 litres.
Drawing a graph through the origin and the point (10, 45·5)
gives us a ready-reckoner.

a Use the graph to estimate how many litres are in:
   i 11 gallons
   ii 8 gallons
   iii 12·5 gallons.

b Use the graph to estimate how many
gallons are in:
   i 30 litres
   ii 55 litres
   iii 15 litres.

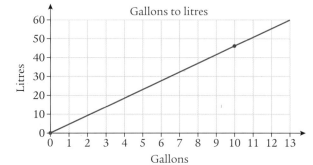

Gallons to litres

Litres / Gallons

3 Eilidh was training for a mountain biking
competition.
The graph shows her progress from the start
of the run to her return.

a How far from the start was she after:
   i 2 hours   **ii** $3\frac{1}{2}$ hours?

b When was she furthest from the start?

c How many kilometres did she cycle altogether?

d Calculate the speed for each section of the run.

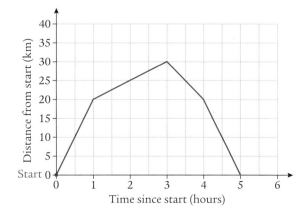

Distance from start (km) / Time since start (hours)

4

Warm Chicken Salad (serves 4)

250 g	baby new potatoes, halved
8	rashers streaky bacon (rinds removed), cut into 2 cm squares
220 g	plum tomatoes
3 tbsp	olive oil
1 tbsp	balsamic vinegar
1	iceberg lettuce heart, leaves washed and separated
2	ripe avocados, peeled and sliced
2	cooked chicken breasts
1	handful fresh basil leaves

a You wish to make this salad for eight people rather than four.

How many rashers of bacon do you need?

b What weight of potatoes would be needed if you were serving six people?

c If you are making this for no people you would need no potatoes (0, 0).

If you are making this for four people you would need 250 g of potatoes (4, 250).

Using the points (4, 250) and (0, 0), draw a graph of 'Number of people' against 'Potatoes (g)'. Let the *x*-axis on the graph run between 0 and 10 people.

d From the graph what weight of potatoes are required for:
   i 3 people   **ii** 10 people?

e If you only have 400 g of potatoes, what is the maximum number of people you can cook for?

Exercise 3.2B

1 The boiling point of water can be written as 100 °C or 212 °F.

The melting point of ice can be written as 0 °C or 32 °F.

a Use this information to create a conversion graph between Celsius and Fahrenheit.

b Use the graph to convert:
   i 30 °C to °F   **ii** 150 °F to °C.

c Calculate the gradient of the line.

2 Divers need to know how much air they are going to use while diving.

The breathing rate in litres per minute increases the deeper you go.

The graph represents the breathing rate against depth below the surface.

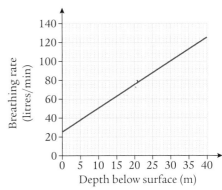

a What is the breathing rate if you are:
   i 20 m   **ii** 40 m below the surface?

b What does the point (0, 25) represent in this situation?

c Calculate the gradient of the line (change in breathing rate per metre).

d How much air, in litres, would a diver use if they were at 10 m for 20 minutes?

51

3 A car hire company charges £25 plus £15 for every day you hire the car.
A table of charges has been started ...

Number of days	0	1	2	3	4	5	6	7
Cost (£)	25	40	55	70				

  a Use the data to draw a line graph.

  b From the graph determine how much it costs to hire a car for seven days.

  c Calculate the gradient of this line.

  d What does the gradient represent in this situation?

  e You have £80 to spend on hiring a car.
  What is the maximum number of days for which you can hire one?

4 John bought a sofa bed for his new house costing £800.
He paid a deposit and then equal monthly instalments
for three years.
The points on the graph (red) show how much was still
owed at the end of each month.
The points all lie on a straight line marked in blue.

  a What was the original deposit?

  b **i** Calculate the gradient of the line.
  ii What is this measurement telling you?

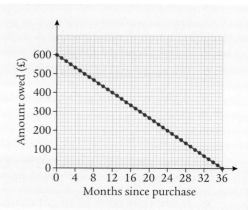

3.3 More real-life graphs

Example 1

A plumber has a call-out charge of £20.
He then charges a further £15 an hour.
Let h hours represent the number of hours for which the plumber works.
The cost, £C, can be expressed in terms of h hours, by the formula:

  $C = 15h + 20$.

a Copy and complete this table of values:

Number of hours worked, h (hours)	1	2	3	4	5
Cost, C (£)	35	50			

b Plot the points and hence draw a line to represent the relationship between C and h.

c Calculate, for this line,
  i the gradient   **ii** the y-intercept where the line crosses the y-axis.

d **i** What does the gradient represent in this context?
  ii What does the y-intercept represent?

e Use the graph to find when the charge first goes over £100.

a

Number of hours worked, h (hours)	1	2	3	4	5
Cost, C (£)	35	50	65	80	95

b

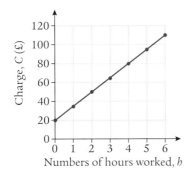

c **i** Taking any two points on the line, e.g. (0, 20) and (1, 35),

$$m = \frac{\text{change in } y\text{-direction}}{\text{change in } x\text{-direction}} = \frac{35 - 20}{1 - 0} = \frac{15}{1} = 15.$$

The gradient is 15.

ii The line cuts the y-axis at 20.

d **i** The gradient represents the hourly rate.

ii The y-intercept represents the call-out charge.

e From the graph, the charge goes over £100 at around $5\frac{1}{3}$ hours.

Exercise 3.3A

1 Hamisi walks 8 km in 90 minutes at a steady pace.

   a By considering the points (0, 0) and (90, 8), draw a distance–time graph to illustrate his walk.

   b Use your graph to estimate:
     i how far he walked in 2 hours
     ii how long he takes to walk 5 km.

2 Shannon cycles from home to the bike shop and back.

   The shop is 6 km away. It takes her 40 minutes to get there.

   She spends 10 minutes at the shop. The return journey takes 30 minutes.

   a Draw a distance–time graph to represent this information.

   b How much faster was she returning home than going to the shop?

     Give your answer in km/h.

3 Sandra earns £6·50 an hour.

   This can be written as $y = 6.5x$, where x represents the number of hours and y represents how much Sandra would earn.

   a **i** Complete a table of values, using even values of x between 0 and 10.
     ii Draw a line graph to illustrate Sandra's earnings.

   b Calculate the gradient of the line. Comment.

4 On a certain date the exchange rate for pounds to dollars was $1·50 per £1.

The relation can be written as $y = 1.5x$, where x is the number of pounds and y is the number of dollars.

  a Complete a table of values, using 0, 100, 200, 300 and 400 as values of x.

  b Hence draw the line graph representing the currency conversion.

  c You want to buy this computer tablet. Use your graph to work out which is the cheaper option.

EMBRA ELECTRICS
Computer Tablet
£370

N.Y.P.C. FUSE
Computer Tablet
$550

5 A hotel makes an initial charge of £100 and then £5 per person for a function room and buffet.

If £C is the total cost and n is the number of people attending the function, the cost of hiring the room and the buffet is given by the formula $C = 5n + 100$.

  a Copy and complete the table of values.

Number of people, n	0	10	20	30	40
Total cost, C (£)	100	150			
Point (n, C)	(0, 100)	(10, 150)			

  b Plot the points to represent the relationship between C and n.

Note that, since you can't get a fraction of a person attending the buffet, the series of points represents the relationship. A straight line can be drawn through them if it is desired that we emphasise that the points lie in a straight line.

  c **i** Calculate the gradient of the line.
    ii Identify the y-intercept.

  d What do:
    i the gradient
    ii the y-intercept
      represent in this situation?

6 The cost of hiring a bicycle is given by the formula $C = 2d + 10$, where £C represents the cost of hiring the bicycle for d days.

  a Complete a table of values, using values for d from 0 to 10 days.

  b Draw a graph to represent this formula.   (Hint: can d be a fraction?)

  c The points lie in a straight line.
    i Calculate the gradient.
    ii State the y-intercept.

  d What do the gradient and y-intercept represent in this situation?

Comparisons

It is often useful to use straight-line graphs to compare two different situations, e.g. tariffs, buying price and selling price, etc. Often, important critical points will become evident.

Example 2

Two mobile phone companies offer different rates for the number of minutes used for a phone call.

Four Mobile Phone Company charges £11 a month plus 15p per minute.

Phone-on-the-Go Mobile Phone Company have no monthly charge and a 20p per minute call rate.

a Copy and complete the table.

Calls per month, N (minutes)	0	20	40	60	80	100
'Four' costs, F (£)	11	14				
'Phone-on-the-Go' costs, P (£)	0	4				

b Draw graphs on the same coordinate diagram to illustrate the data.

c Where do the lines cross?

d Which company would be the cheaper if you made:

   i 140 minutes   **ii** 260 minutes

      of calls a month?

e Write down an equation in terms of N for:

   i F   **ii** P.

a

Calls per month, N (minutes)	0	20	40	60	80	100
'Four' costs, F (£)	11	14	17	20	23	26
'Phone-on-the-Go' costs, P (£)	0	4	8	12	16	20

b

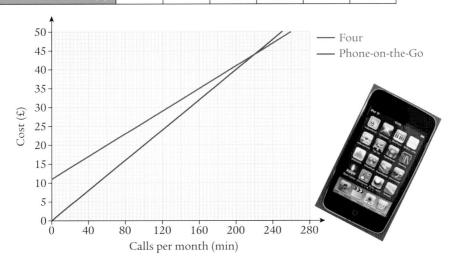

c The lines cross at (220, 44).

d **i** At 140 minutes, Phone-on-the-Go is cheaper (red line below blue).

   ii At 260 minutes, Four is cheaper (red line above blue).

e **i** $F = 0.15N + 11$ **ii** $P = 0.2N$

Exercise 3.3B

1   While on holiday in Tenerife you can hire a scooter to travel round the island.

There are two different rental companies close to your resort:
- Rent-a-Bike charges €20 a day plus 25 cents a kilometre.
- Scooter-to-Go charges €55 a day with no distance charge.

a   Copy and complete the table of values.

Distance, d (km)	0	20	40	60	80
Rent-a-Bike, R (€)	20	25			
Scooter-to-Go, S (€)	55	55			

b   Draw line graphs on the same diagram to illustrate the data.

c   Where do the lines cross?

d   Which company would be the cheaper if you travelled:
  i 90 kilometres   **ii** 120 kilometres?

e   At which point do both companies charge the same?

f   Write down an equation in terms of d for: **i** R **ii** S.

2   A heating engineer charges an hourly rate plus a fixed fee for coming out (a call-out charge).
On a weekday during normal working hours he asks for £25 call-out and then £9 an hour.
He'll only charge a fraction of the £9 for a fraction of an hour, e.g. £3 for 20 minutes.

a   Make a table showing his total charges for jobs between one hour and five hours.

b   Draw a graph to represent the data.

c   When it is not normal working hours he charges more ... £40 call-out and £12 an hour.
On the same diagram, draw a line to represent these dearer charges.

d   For each line find: **i** the gradient   **ii** the y-intercept.

e   What do the gradient and y-intercept mean in each of these situations?

3   When running a business, the **break-even** point is when the income is equal to the expenditure.
This can easily be found by representing income and expenditure on a graph.
Jane started a cupcake business.

Her expenditure:
There is a fixed cost of £12 for equipment and a cost of £0·75 per cupcake.

a   Draw a line graph showing her costs for 0 up to 25 cupcakes.

Her **income:** She sells her cakes at £1·50 per cupcake.

b   On the same diagram, draw a line graph of income made
on 0 up to 25 cupcakes.

c   **i** At what point do the lines cross? (This is the break-even point.)
  ii How many cupcakes does she need to sell to break even?

d   What profit does she make after selling 30 cupcakes?

4 Simon has started creating and selling mobile phone apps.
It costs Simon £20 for materials. It also costs him £0·50 per app.

   a Draw a line graph to show how the total cost changes the more apps he creates.

   b He sells his apps at £2·50 per app.
Add a line graph showing his income as the number of apps sold increases.

   c Identify the break-even point.

   d What profit does he make after selling 30 apps?

Preparation for assessment

1 Calculate the gradient of each of the following slopes:

a   **b**   **c**

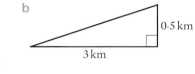
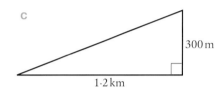

30 m   0·5 km   3 km   300 m   1·2 km   24 m

2 The gradient of a road is given as 15%.

   a Write this as: **i** a common fraction   **ii** a ratio.

   b What would be the vertical rise of the road over a horizontal distance of 0·8 km?

3 a There are two routes to the top of Ben Dearg.
The route from A to X via B and C covers a distance of 1000 m.
AB is 180 m, BC is 270 m and CX is 550 m.
Calculate the gradient for each of these sections.

   b The route from D to X via E and F is only 550 m long.
DE is 250 m, EF is 150 m and FX is 150 m.
Calculate the gradient for each section of this route.

   c Calculate the average gradient from:
    i A to X   **ii** D to X.

   d Describe one benefit and one drawback of each route.

4 An electrician charges £25 on call-out and then a further £5 an hour till the job is done.
He doesn't call fractions of an hour a whole hour.
The cost, £C, can be calculated using the formula $C = 5h + 25$, where h hours is the time spent on the job.

   a Copy and complete this table of values.

Hours worked, h (hours)	0	1	2	3	4	5
Cost, C (£)	25	30	35			

   b Draw a line graph to represent the relationship between C and d.

57

    c Calculate: i the gradient   ii the *y*-intercept of the straight line.

    d   i What does the gradient represent in the context?

         ii What does the *y*-intercept represent in the context?

5 There is a relationship between the temperature in degrees Celsius and the number of chirps a cricket makes in 25 seconds!

   It only works between 13 °C and 38 °C.

   Outside this range the cricket is too cold or too hot to sing.

   The graph illustrates the relationship.

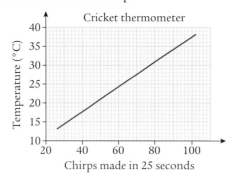

    a Estimate the temperature when the cricket makes: i 30 chirps   ii 90 chirps in 25 seconds.

    b How many chirps would you expect to hear if the temperature were 19 °C?

6 Jonathon was training for a marathon. The graph shows one of his training runs.

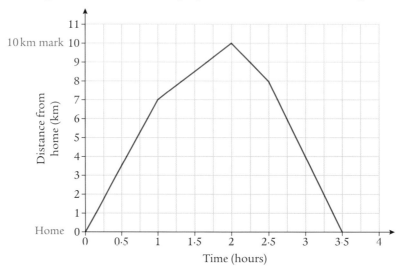

    a In the first leg he covered 7 km in 1 hour, moving away from home.

       Describe the other three legs in a similar way.

    b How many kilometres did he run altogether?

    c Calculate the gradient, and hence his speed, for each leg of the run.

7 Tim and Karen are looking for a baby-sitter.

   One person, Peter Piper, charges a £1 booking fee and £3 an hour after that.

   Mary Merrie charges a £3 booking fee but only asks for £2 an hour.

Below is an incomplete graph showing both sets of fees for up to four hours.

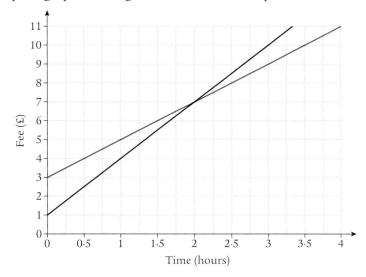

a   Which line represents the fees for which sitter?

b   For which length of time will both sitters charge the same amount?

c   Tim and Karen need a sitter for three hours. Which of the sitters offers the cheaper deal?

8   Remember the holiday? You are going to Australia and need to change some money.
Different companies charge different amounts for the service.

Holiday Cash
RATE: £1 BUYS AUS $1·60
COMMISSION: 10%

Xchange
RATE: £1 BUYS AUS $1·50
COMMISSION: £5 PER TRANSACTION

COMMISSION
CHARGED BEFORE
EXCHANGE OCCURS

Which is the best company for you?

Investigate the situation.

Before we start...

Chris wants to buy a car.

He has saved some money but still needs to borrow £2500.

Chris has two choices: get a loan or use a credit card.

<u>Blue Sky Credit Card</u>

APR 15·9%

BANK LOANS

Borrow £1000
Repayment Scheme
12 monthly payments of £93·02

Which choice would give Chris the better deal?
Justify your answer.

What you need to know

1 Round the following to 2 decimal places:

  a 4·568   b 13·083   c 25·2346   d 51·89923.

2 Write the following percentages as decimal fractions:

  a 3%   b 18%   c 8·3%   d 1·25%.

3 Calculate the following, giving your answer correct to 2 decimal places:

  a 4% of £14·86   b 17% of £67·56

  c 1·4% of £15·30   d 2·56% of £654·89.

4 a Increase £500 by 15%.   b Decrease £30 by 10%.

5 During a sale a car's price was dropped from £25 000 to £20 000.

  Express the drop as a percentage of the original price.

4.1 Currencies

Foreign exchange

When you travel abroad you will need to exchange your British money for the local currency.

Exchange rates tell you how many units of one currency you can get of another currency.

Exchange rates vary from day to day and different vendors (banks, building societies, travel agents, post offices and supermarkets) will offer different rates.

Some vendors also charge a fee for exchanging money, called **commission**.

When changing money, you should shop around for the best deal.

The table below shows the exchange rates for some countries.
(These exchange rates were correct in December 2012.)

Country	Currency (unit)	Exchange rate (units per £1)
Eurozone	euro (€)	1·181
USA	dollar ($)	1·563
Turkey	Turkish lira	2·726
Hong Kong	Hong Kong dollar (HKD)	11·72
South Africa	rand	13·05
Poland	zloty	4·692

If you're in Britain when you buy euros for example, the exchange rate will be given as euros per pound.

If you are somewhere in the Eurozone and wish to change some more British money, the exchange rate will be given as pounds per euro.

Example 1

Jenny is going to Germany on holiday. She wants to change £250 into euros.
Using the above exchange rates, how many euros will she receive?

For £1 she will get €1·181   From the table
So for £250 she will get 250 × €1·181 = €295·25.

Example 2

David buys a book for £6·99.

When visiting his uncle in New York, he sees the same book on sale for $9·95.

How much cheaper is the book in New York? (Give your answer in pounds sterling correct to 2 decimal places.)

$1·563 = £1
⇒ $1 = £1 ÷ 1·563 = £0·639795...   Don't round if you're
⇒ $9·95 = 9·95 × £0·639795... = £6·37 (to 2 d.p.)   using a calculator

Difference in cost = £(6·99 − 6·37) = £0·62.

The book is cheaper in New York by £0·62 = 62p (to 2 d.p.).

Example 3

Margaret is in Rome and runs out of local currency.

She goes to the bank to change £200 into euros.

The sign in the bank says, '£0·85 per euro'.

How many euros will she get?

£0·85 = €1

$\Rightarrow$ £1 = €1 ÷ 0·85 = €1·17647059...

$\Rightarrow$ £200 = 200 × €1·17647059...

       = €235·29 (to 2 d.p.)

Margaret will get €235·29.

Exercise 4.1A

Unless otherwise stated, use the exchange rates in the above table.

1 **a** Change £50 to euros. **b** Change £200 to Hong Kong dollars.

   c Change £1400 to zloty.

2 Katherine is going to France on holiday and wants to change £320 to euros.

   a How many euros will she receive?

   b Having spent €350, she changed the rest back into pounds.
   The bank will only exchange whole numbers of euros.
   At the same exchange rate,
     i how many pounds does she get
     ii how many cents is she left with?

3 A small mail order company accepts all currencies in payment.
   Three payments arrive: $65, 1150 Turkish lira, 568 rand.

   a Change each amount to British pounds.

   b The bank takes a 2% commission for handling the transaction.
   How many pounds does the company end up with?

4 Chanda bought some presents for her family when travelling around Europe.
   Change the price of each present into pounds, correct to the nearest penny.

Earrings €25   Bag €8·95   Fridge magnet €3·50

5 Whilst in Florida, Jardine buys a pair of trainers for $85.

When she returns to Edinburgh, she finds the same pair of trainers for £59·99.

How much money did she save by buying the trainers on holiday?

6 Michael went to Bruges for a fortnight.

He paid €60 a night for his hotel for 13 nights.
He paid €40 a day for food for 14 days.

He started his holiday with £2000, all of which he converted to euros.

a After paying for food and the hotel, how much did he have left for other things?

b He changed another £150 into euros while he was there.

He was given a rate of £0·88 to the euro.
How many euros did he get?

c When he went home he had €52·76 left.

The bank turned them back into pounds, accepting whole euros only.

They quoted a rate of €1·15 to the pound.
How many pounds did he get?

Example 4

Marsha came over from Boston to visit Europe. She intended to stop first in Edinburgh.

She changed $2000 at a rate of £0·65 to the $1.

a How many pounds sterling did she get?

b Having spent £800, she converted the remainder to euros, getting €1·20 to the pound.

How many euros did she get?

c She spent €550 and converted the rest to dollars to go home. The rate quoted was €0·76 per dollar. How many dollars did she get?

a $1 = £0·65

⇒ $2000 = 2000 × £0·65 = £1300

b £1300 − £800 = £500

£1 = €1·20

⇒ £500 = 500 × €1·20 = €600

c €600 – €550 = €50

€0·76 = $1

⇒ €1 = $1 ÷ 0·76 = $1·315789...

⇒ €50 = 50 × $1·315789...

= $65·79 (to 2 d.p.)

Marsha got $65·79 to take home.

Example 5

At the post office Michael noticed that he could get €1·20 to the pound (£).
He also saw that he could get 13 South African rand to the pound (£).

a How many rand are there to the euro?   **b** How many euros do you expect to get for 200 rand?

a €1·20 = 13 rand.

$\Rightarrow$ €1 = (13 ÷ 1·20) rand
= 10·83 rand (to 2 d.p.)
€1 = 10·83 rand.

b 13 rand = €1·20.

$\Rightarrow$ 1 rand = €1·20 ÷ 13
= €0·0923...
$\Rightarrow$ 200 rand = 200 × €0·0923...
= €18·46 (to 2 d.p.).

Exercise 4.1B

Refer to the table below for exchange rates.

Country	Currency (unit)	Exchange rate (units per £1)
Eurozone	euro (€)	1·181
USA	dollar ($)	1·563
Turkey	Turkish lira	2·726
Hong Kong	Hong Kong dollar (HKD)	11·72
South Africa	rand	13·05
Poland	zloty	4·692

1 Stephen receives €50 for his birthday from his sister who lives in Spain.
He wants to buy a watch costing £39·99. Can Stephen afford to buy the watch?

2 **a** Holly wants to change £1000 to euros for a trip to Austria.
How many euros will she get?

b When Holly is in Austria, she decides to go to Budapest, in Hungary, for a day.
She wants to change €90 to forint. The exchange rate is £1 = 320 forint.
How many forint will she get?

3 Andrea wants to change £150 into dollars ($). Her bank charges a commission of 2%.

a Calculate the commission charged.

b The commission is deducted before the money is changed.
How many dollars will Andrea obtain?

4 A supermarket sells euros for Australian dollars at a rate of €1 = 1·25 Australian dollars.

a Priya wishes to exchange €550 for Australian dollars.
How many dollars will she receive?

b Priya spends 463·25 Australian dollars.
She then sells her remaining dollars back to the supermarket at a different rate
of €1 = 1·38 Australian dollars.
How many euros does Priya have remaining?

5   Trevor exchanges £680 for 6230 Danish kroner (singular: krone).
    Calculate the exchange rate that was used to convert:

    a   pounds to kroner   b   kroner to pounds.

6   At the travel agent, Kimberley noticed that she could get €1·21 to the pound (£).
    She also saw that she could get 88 Indian rupees to the pound (£).

    a   How many Indian rupees are there to the euro?

    b   How many euros do you expect to get for 5000 rupees?

7   When travelling abroad there are many different options for travel money: foreign currency,
    traveller's cheques, debit cards, credit cards and pre-paid cards.
    Investigate the benefits and disadvantages of these options.
    Consider safeguards and insurance.

4.2 Savings

When you put money in a bank or building society, it earns extra money called **interest**.

The money deposited in the bank to start with is called the **principal** amount.

After that, as your money 'grows', what is in the bank is referred to as the **amount**.

The interest earned is a percentage of what is in the bank.

The **rate of interest** is set by the bank and differs for different schemes.

It is commonly quoted for a year, e.g. 3% per year or 3% p.a.

Example 1

Joanna deposits £850 into a bank account with an interest rate of 2% per annum.

a   Calculate the interest earned after one year.

b   What is the amount in Joanna's account after one year?

a   The interest is 2% of £850 $= \dfrac{2}{100} \times £850 = £17.$

    Joanna earns £17 interest after one year.

b   £850 + £17 = £867.

    After one year, Joanna has £867 in her account.

Simple interest

When the interest earned is calculated on the principal amount, it is called **simple interest**.

If the money is left in the bank for, say, five years, then five lots of the same interest are earned.

p.a. stands for *per annum* and means 'per year'.

Example 2

The ABC Building Society offers simple interest at an introductory rate of 1·2% p.a. for new accounts opened.

If Joshua opens an account with £1320, how much money will he have after four years?

Interest after one year: 1·2% of £1320 $= \dfrac{1·2}{100} \times £1320 = £15·84$.

Interest after four years: £15·84 × 4 = £63·36.

Total amount in account after four years:
£1320 + £63·36 = £1383·36.

Compound interest

When the interest earned is calculated on the amount in the account, it is called **compound interest**.

Each year, when the interest is calculated and added to the account, the amount grows ... and the following year's interest is calculated on a bigger amount.

Example 3

The ABC Building Society offers **compound interest** at a rate of 1·2% p.a. in a special saver's account.

If Joshua had put his £1320 into this account,

a how much money would he have had in his account after four years?

b what interest would this have earned over the four years?

a Note that 1·2% $= \dfrac{1·2}{100} = 0·012$.

Interest at end of year 1 = 0·012 × 1320
= £15·84.

Amount at end of year 1 = 1320 + 15·84
= £1335·84.

Interest at end of year 2 = 0·012 × 1335·84
= £16·03.

Amount at end of year 2 = 1335·84 + 16·03
= £1351·87.

Interest at end of year 3 = 0·012 × 1351·87
= £16·22.

Amount at end of year 3 = 1351·87 + 16·22
= £1368·09.

Interest at end of year 4 = 0·012 × 1368·09
= £16·42.

Amount at end of year 4 = 1368·09 + 16·42
= £1384·51.

So the amount in Joshua's account after four years = £1384·51.

b Interest earned over four years = £1384·51 − £1320 = £64·51.

Exercise 4.2A

1 Calculate the interest gained after a year on the following amounts at the given rates:

a £1200 at 1% p.a.

b £620 at 4% p.a.

c £75 at 2% p.a.

d £810 at 1·3% p.a.

e £2400 at 0·5% p.a.

f £5630 at 0·8% p.a.

2 Roger invests £400 into the Platinum Account.

   a How much interest will he earn after one year?

   b How much money will Roger have in the account after one year?

PLATINUM ACCOUNT
Interest: 1.5% p.a.

3 Toby is given £90 for his birthday.

He decides to spend one third of his money.

   a How much money does Toby spend?

   b Toby then saves the rest in a bank account with 0·8% interest per annum.

     i How much interest will Toby receive after one year?

     ii If Toby had decided to save all of his birthday money, how much extra interest would he have earned after one year?

4 Alice puts £1600 into a bank account paying 1·65% interest p.a.

How much interest will Alice get after five years:

   a if it is simple interest   **b** if it is compound interest?

5 Ingrid puts £1240 into a bank account paying 1·04% interest p.a.

How much money will Ingrid have in her account after three years:

   a if it is simple interest   **b** if it is compound interest?

6 Karen turns 18. She is given £3000, which she cannot withdraw till she's 21.

So she invests it in a scheme for three years that offers 3% compound interest p.a.

   a How much will be in the account after three years?

   b How much interest has it earned in three years?

Parts of a year

In the following section, money left in the bank for a fraction of a year will attract that fraction of a year's interest. This is not always the case. Sometimes the special rate depends on keeping your money locked away for a fixed time.

Example 4

Penny invests £2650 into the account offered by the Royal Bank.

a How much interest will Penny receive after one year?

b Penny decides to withdraw her money after eight months. Calculate how much interest she will get now.

ROYAL BANK

Special offer: 2·5% interest per annum

a Interest after one year: 2·5% of £2650 = $\frac{2 \cdot 5}{100} \times £2650 = £66 \cdot 25$.

b Interest after one month: $\frac{£66 \cdot 25}{12} = £5.520833... = £5 \cdot 52$ (to 2 d.p.).

Interest after eight months: £5·52 × 8 = £44·16.

(Note: banks generally round down but the real situation may vary when it comes to the practice of rounding.)

Exercise 4.2B

Round answers to 2 decimal places.

1 Calculate the interest earned on:
  a £400 at a rate of 3% p.a. for 7 months
  b £8000 at a rate of 1% p.a. for 3 months
  c £105 000 at a rate of 0·75% p.a. for 11 months.

2 The Super Saver Account offers an interest rate of 1·4% per annum.
  Carrie puts £750 into the Super Saver Account.
  How much interest will Carrie receive after:
  a 1 month
  b 5 months?

3 The Swift Building Society has a savings account with a rate of interest of 1·75% p.a.
  Tim and Lucy invest £1500 and £1300 in this account respectively.
  Tim withdraws his money after four months.
  a How much interest will Tim earn?
  b Lucy takes out her money after six months.
    Who earned more interest and by how much?

4 Eliza puts £30 into a building society account with a rate of interest of 2·5% p.a.
  How many months will it take Eliza to earn £6 in interest?

5 Alexander is choosing between two different bank accounts.
  Both offer compound interest.

Steady Saver
Fixed rate: 1·7% p.a.

Flexible Investment

3% p.a. for the first year
1% p.a. thereafter

He intends to invest £1500 for five years.
Which account should he choose? Justify your answer.

6 Research the range of savings accounts offered by different financial organisations.

Look at: rate of interest, tax status, conditions, withdrawal arrangements, bonuses.

7 Kate, Stuart and Zoe each want to open a savings account.

The bank offers three different savings accounts.

Read the information about the different accounts and decide which account is best suited for Kate, Stuart and Zoe.

Sensible Saver

- 0·9% rate of interest per annum
- Cash card for instant access
- Minimum balance of £1

Kate is in 4th year at high school at the moment. She is about to start a part-time job and wants to save for when she goes to university.

Web Saver

- 1·4% rate of interest per annum
- Online access only
- Minimum balance of £1

Stuart has recently bought a house and wants to set up a savings account for household emergencies.

Ultimate Saver

- 2·6% rate of interest per annum
- Make withdrawals on three days only each year
- Minimum balance of £100

Zoe has been given money for her birthday and wants to save it. However, she knows she will want to take money out regularly to go shopping.

4.3 Borrowing – loans

Loans

When you save money in a bank or building society, you earn interest.

So when you borrow money, you are expected to pay interest.

When you borrow money from a financial organisation, you need to know:

- the monthly rate of interest
- the APR
- how long you have got to repay the loan.

Payment Protection Insurance (PPI) can be purchased with a loan to help cover part of the cost of a loan should the borrower become unable to pay back the loan because of illness, redundancy or other circumstances.

APR (Annual Percentage Rate) is the interest rate for a whole year.

Example 1

Mr Watson takes out a loan of £1000 from the Barnes Building Society.

He has to pay £19·04 per month for 60 months.

a Calculate how much Mr Watson will repay in total.

b What is the cost of the loan?

a Total repayment: 60 × £19·04 = £1142·40.

Mr Watson will have to pay £1142·40 back to the Building Society.

b £1142·40 − £1000 = £142·40.

The loan will cost Mr Watson £142·40.

Example 2

Anna wants to borrow £3500.

Calculate her monthly repayment.

> **Borrow £1000**
> **Payback over 5 years.**
> *60 monthly intalments of £ 50·25*

First calculate how many £1000s are borrowed:

$$\frac{3500}{1000} = 3.5.$$

It costs £50·25 a month per £1000 borrowed.

Therefore, for a £3500 loan, it will cost 3·5 × £50·25 = £175·88 a month.

Exercise 4.3A

The table shows the monthly repayments on a loan of £1000.

APR (%)	Loan term (months)				
	12	18	24	36	48
5	£85·60	£57·80	£43·90	£30·00	£23·10
6	£86·10	£58·20	£44·30	£30·40	£23·50
7	£86·50	£58·70	£44·80	£30·90	£23·95
8	£87·00	£59·10	£45·20	£31·30	£24·40

1 For each of the following loans, calculate:

  i the monthly repayment

  ii the total repayment

  iii the cost of the loan.

  a £1000 borrowed for 12 months with APR of 7%.

  b £1000 borrowed for 36 months with APR of 8%.

  c £1000 borrowed for 24 months with APR of 6%.

  d £2000 borrowed for 18 months with APR of 5%.

  e £4000 borrowed for 48 months with APR of 6%.

2 Fred Brown takes out a loan of £6000 to buy a new kitchen.

He will repay the loan over 36 months with an APR of 6%.

  a How many thousands has Fred borrowed?

  b Calculate his monthly repayment.

  c Find the total amount repayable.

  d What does the loan cost Fred?

3 Stacey wants to buy a new car.
She borrows £3000 at a rate of 8% for a period of 18 months.

a How many thousands has Stacey borrowed?   **b** Calculate her monthly repayment.

c Find the total amount repayable.   **d** What does the loan cost Stacey?

4 Jimmy borrows £7500 to pay for a new conservatory.
He will pay back the loan over 48 months with an APR of 5%.

a How many thousands has Jimmy borrowed?

b Calculate Jimmy's monthly repayment.

c Find the total amount repayable.

d What does the loan cost Jimmy?

5 The table below shows the monthly repayments on a loan of £1000, to be paid over 60 months, from different lenders.

a For each lender, calculate:
  i the total repayment
  ii the cost of the loan.

b Which lender is offering the cheapest loan?

Lender	Monthly repayments
East Bank	£19·35
Rose Building Society	£19·49
Hill Bank	£19·27
Swift Building Society	£19·04

Example 3

Mr O'Donnell takes out a loan for £3000.
He is charged simple interest at the rate of 9·4% p.a.
He agrees to make equal monthly payments until the debt is paid off.

a **i** Calculate the total amount to be paid back if he takes a year to pay it.
  ii Calculate the size of each monthly repayment.

b Mr O'Donnell decides to repay the loan over six months.
  i Calculate the interest for six months.
  ii What is the total amount to be paid back?
  iii What would his monthly repayments be?

a **i** Annual interest: 9·4% of £3000 $= \dfrac{9\cdot4}{100} \times £3000 = £282$.

  Total to repay: £3000 + £282 = £3282.

  ii Monthly payment: $\dfrac{£3282}{12} = £273\cdot50$.

b **i** Annual interest: £282.

  Interest for six months: $\dfrac{£282}{2} = £141$.

  ii Total to repay: £3000 + £141 = £3141.

  iii Monthly payment: $\dfrac{£3141}{6} = £523\cdot50$.

Example 4

£1000 is borrowed by Malcolm to buy a smart TV.

At the end of each month 1% interest is added to the debt and Malcolm pays a fixed £200 until the debt is less than £200. His final payment is enough to clear the debt.

a What was the size of his final payment?

b What did he pay in total?

a End of 1st month: £1000 + (0·01 × £1000) − £200 = £810.
End of 2nd month: £810 + (0·01 × £810) − £200 = £618·10.
End of 3rd month: £618·10 + (0·01 × £618·10) − £200 = £424·28.
End of 4th month: £424·28 + (0·01 × £424·28) − £200 = £228·52.
End of 5th month: £228·52 + (0·01 × £228·52) − £200 = £30·81.
End of 6th month: £30·81 + (0·01 × £30·81) = £31·12.
Final payment is £31·12.

b Total payment = 5 × £200 + £31·12 = £1031·12.

Exercise 4.3B

1 Ursula borrows £8300.
She agrees to an APR of 7%.
She pays the debt back with 12 equal monthly instalments.
  a Find the total amount repayable.   **b** What does the loan cost Ursula?
  c What is the size of each instalment?

2 Zach decides to borrow £2250 to buy a boat.
He opts for a rate of 8%. He agrees to pay the debt with 24 equal monthly instalments.
  a Find the total amount repayable.   **b** What does the loan cost Zach?
  c What is the size of each instalment?

3 Miss Preston takes out a loan of £1000.
She is charged simple interest at the rate of 8·7% p.a.
She agrees to make equal monthly payments until the debt is paid.
  a Calculate the monthly payments if she pays over 12 months.
  b She decides to repay the loan over six months. What would her monthly payments be?

4 Kayleigh takes out a loan of £6000.
Each month, interest at the rate of 1·2% is added to the debt.
She then makes a repayment. She's told the minimum repayment is £72 a month.
Thinking to clear the debt after 12 months, she agrees to pay £500 a month.
  a Calculate the size of the debt after the fourth payment.
  b She decides to repay the loan over three months.

If she makes monthly repayments of £2048, how much is left to be paid at the end of the third payment?

c Comment on what would happen if she only made the minimum repayment.

5 Mr and Mrs George are going to get a loan of £2000 from their bank.

They are given two options.

Option	1	2
Loan term (months)	12	4
APR (%)	10·5	8·6

a Calculate the monthly payments for each option.

b Which loan should Mr and Mrs George take:
  i if they want the cheapest loan
  ii if they can only afford a monthly repayment of at most £300?
  Show your working.

6 Mr and Mrs George decide to shop around and consider two deals from other banks.

Option	3	4
Loan term (months)	6	9
APR (%)	9·9	9·1

a Calculate the monthly payments for each option.

b Should Mr and Mrs George select one of these options:
  i if they want the cheapest loan
  ii if they can only afford a monthly repayment of at most £300?
  Give a reason for your answers.

7 Jeremy takes out a loan for £7000.

He pays it back over four years at an annual percentage rate of 3·2%.

When the debt falls below £1, it is cancelled.

a Show that an annual payment of £1892, made after the interest is added, is enough to have the debt cancelled by the end of the fourth year.

b He purchases PPI at a price of 16% of his initial loan.

  Calculate: i the PPI charge   ii the total cost of the loan.

8 Hilary wants to apply for a loan of £12 000.

The monthly payment is £380·25 and the loan has to be repaid over 60 months.

a Calculate the total amount to be repaid.

b Hilary is offered the option of buying PPI at a cost of £38 per month.

  How much extra will she pay in total if she purchases the PPI?

c What is the new total now to be repaid?

d What will the loan cost Hilary?

4.4 Borrowing – credit cards and store cards

Credit cards and store cards

A **credit card** is a way of borrowing money.

You can buy items and pay for them later, usually at the end of each month.

You will receive a monthly statement showing how much you have spent, the amount you owe and the minimum amount you have to pay back that month.

Any amount that is left unpaid at the end of the month has interest added to it.

Just as with savings, the interest is calculated as a percentage.

A **store card** is a form of credit card used to buy items from that store only.

As with credit cards, you receive a monthly statement and any amount that is left unpaid at the end of the month has interest added to it.

Store cards often have a higher rate of interest than a credit card.

Note on APR

When these cards are advertised, both the monthly rate of interest and the APR (Annual Percentage Rate) is given. Note that the APR is not simply 12 times the monthly rate.

The table shows what some of the monthly rates become when expressed as an APR.

What looks all right as a monthly rate can be quite frightening when expressed as an APR.

A monthly interest rate of 6% means an annual percentage rate of over 100%.

Monthly rate (%)	APR (%)
1.0	12.7
1.1	14.0
1.2	15.4
1.3	16.8
1.4	18.2
1.5	19.6
1.6	21.0
1.7	22.4
1.8	23.9
1.9	25.3
2.0	26.8

Example 1

Annie takes out a Lion Credit Card with monthly rate 1·52% (APR 19·8%).

a Calculate the interest Annie would owe at the end of the first month if she spent £450 on her credit card.

b How much interest would Annie have to pay in a year on this amount if she didn't pay any of it off?

a Interest is 1·52% of £450 $= \dfrac{1·52}{100} \times £450 = £6·84$.

Annie owes £6·84 interest after the first month.

b The APR is 19·8%.

In a year the interest would be 19·8% of £450 $= \dfrac{19·8}{100} \times £450 = £89·10$.

Annie would have to pay £89·10 interest at the end of the year.

Example 2

Polly's first credit card charges 2·4% interest per month (32·9% APR).

Within one day of getting her credit card, Polly reaches her credit limit by spending £1200.

If Polly does not buy anything else on the card and makes payments of £150 per month, how much money would she owe after two months?

Interest after first month: 2·4% of £1200 $= \dfrac{2·4}{100} \times £1200 = £28·80$.

Balance on card at end of first month: £1200 + £28·80 − £150 = £1078·80.

Interest after second month: 2·4% of £1078·80 $= \dfrac{2·4}{100} \times £1078·80 = £25·89$.

Balance on card at end of second month: £1078·80 + £25·89 − £150 = £954·69.

Polly's debt is £954·69 after two months.

Exercise 4.4A

1   Kyle charges £100 to his credit card which has a 1·4% monthly interest rate.
    a   Use the table to quote the APR.
    b   How much interest will he have to pay at the end of the first month?

2   Felicity spends £250 on her credit card which has a 2·5% monthly rate of interest.
    a   Calculate the interest Felicity will owe after the first month.
    b   How much will she have to pay in total at the end of the month?
    c   The equivalent APR is 34·5%.
        How much interest would be due on £250 if none is paid for a year?

3   Mona's balance on her credit card statement was £247·93.
    If her minimum payment is 3·75% of the balance, what must she pay?

4   On Mario's credit card bill, it says that he is due to make a minimum payment of 2·4% of the balance, or £10, whichever is greater.
    How much will he pay if his balance is £456·80?

5   Brian has a balance of £250 on his credit card at the end of January.
    He has to make a minimum payment of 2·25% of the balance.
    a   Calculate the minimum payment that Brian has to make.
    b   What is the new balance on Brian's credit card now?

6   Frank has a balance of £1250 on his credit card, with a monthly rate of interest of 1·3%.
    a   Calculate the interest that Frank has to pay at the end of the month.
    b   Frank's sister, Wendy, also has a balance of £1250 but her credit card provider charges 1·7% interest per month.
        How much more interest does Wendy have to pay than Frank?

7 Lisa uses a credit card with a monthly rate of interest of 1·7% (APR 22·4%).

    a Calculate the interest Lisa would owe at the end of the first month if she spent £300 on her credit card.

    b How much interest would Lisa have to pay in a year on this amount assuming she did not use her card again and did not pay any of it off?

Exercise 4.4B

1 Muriel takes out a store card for a department store.
She receives a summary of the account.

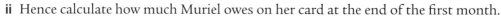

> Monthly rate 1·81% (APR 24·0%)
> Credit limit £250
> Late payment fee £12
> Discount: 15% off all purchases in the first month

    a Muriel spends £120 in the store in the first month.

      i Calculate the discount she will receive.

      ii Hence calculate how much Muriel owes on her card at the end of the first month.

    b Calculate the interest due on this amount at the end of the first month.

    c Muriel makes her first payment of £20 a few days late.

    How much does she owe now, assuming the late payment fee is added after the payment is made?

2 Julia's credit card charges 1·9% interest per month on the balance at the end of a month.

In September, Julia owes £800.

If she makes a payment of £75 each month, how much money would she owe after three months?

Copy and complete:

Interest at end of September: $\dfrac{1·9}{100} \times £800 = £15·20$.

Balance on card at end of September: £800 + £15·20 − £75 = £740·20.

Interest at end of October: $\dfrac{1·9}{100} \times £740·20 = £...$

Balance on card at end of October: £740·20 + ... − £75 = £...

Interest at end of November: $\dfrac{1·9}{100} \times ... = £...$

Balance on card at end of November: ... + ... − £75 = £...

3 In a month Declan spends £1350.

His credit card charges 0·85% interest per month.

He makes a payment of £170 per month.

How much money would he owe after two months?

4 The Silver Credit Card offers 0% interest for the first three months then a monthly interest rate of 1·6%.

Norman spends £560 in the first month, £125 in the second month and £575 in the third month.

He pays back £180 at the end of each month. How much will Norman still have to pay at the end of six months?

5 Create a leaflet explaining about credit cards – what they are, how they work, advantages/disadvantages, how to choose the most suitable card, debt problems, etc.

6 Explore the relation between monthly rate and APR.

You can use a spreadsheet to investigate.

In A1 enter the monthly rate.

In A2 type: =((1+A1/100)^12-1)*100

This will calculate the APR.

One advertisement quoted an APR of 278%!

Use trial and error to find the monthly rate they are charging.

Look at the rates in other adverts for loans and see how debt can grow out of control.

4.5 The best deal

When buying a product it is important to compare prices to make sure you get the best deal. The price of a product may vary from shop to shop and sometimes products are packaged in different sizes.

To find the size that gives the best deal, it is best to calculate the unit price of the contents.

Example 1

Mrs Harrison wants to buy a box of tea bags.

Box A contains 10 tea bags and costs £1·60.

Box B contains 24 tea bags and costs £3·36.

Which is the better buy?

We need to find the price of one tea bag.

Box A: $\dfrac{£1·60}{10} = £0·16$ per bag.

Box B: $\dfrac{£3·36}{24} = £0·14$ per bag.

Box B is the better buy since it costs £0·14 per bag while Box A costs £0·16 per bag.

Example 2

A local supermarket sells bars of chocolate.

There are three choices available:

     option A: 360 g bar of chocolate costs £7·50

     option B: 45 g bar of chocolate costs 68p

     option C: 1 kg bar of chocolate costs £10.

a Which bar of chocolate offers the best value for money?

b Why might someone not go for the 'best buy'?

a Option A: 360 g costs £7·50 $\Rightarrow$ 1 g $= \dfrac{£7·50}{360} = £0·0208...$ (£21 per kg).

Option B: 45 g costs £0·68 $\Rightarrow$ 1 g $= \dfrac{£0·68}{45} = £0·0151...$ (£15 per kg).

Option C: 1 kg costs £10 $\Rightarrow$ 1 g $= \dfrac{£10}{1000} = £0·01...$ (£10 per kg).

Option C offers the best value for money because it is the cheapest price per gram.

b However, 1 kg of chocolate is a large quantity so is only the best deal if you actually want a large amount of chocolate.

Exercise 4.5A

1 Chocolate Whatnots come in two sizes.

Packet A weighs 80 g
and costs £2·40

Packet B weighs 150 g
and costs £3

Weight for weight, which is the better buy?

2 Bags of apples are sold in two different sizes: 6 apples for £2·94 and 10 apples for £5·49.

Assuming the apples themselves are of equal size, which is the better buy?

3 A shop sells three different sizes of jars of coffee.

Jar A
300 g for £4·66

Jar B
100 g for £2·78

Jar C
50 g for £1·89

Weight for weight, which is the best buy?

4 A cereal bar is sold in packs of 12, 24 and 48 at prices of £1·39, £2·39 and £3·94 respectively.

   a Which size of packet offers the best value for money?

   b A special offer on the 24-pack of cereal bars enables you to buy two packs for £4.

      Would you be tempted by this offer? Explain your answer.

5 Susie wants to buy lemonade. There are three different options for her to choose from:

        option A: 8 × 330 ml cans costing £3·85

        option B: 12 × 150 ml cans costing £4·20

        option C: 4 × 2-litre bottles costing £6·49.

   a Which option offers the least value for money?

   b Which option offers the best value for money?

Exercise 4.5B

1 Sarah wants to buy a new phone. She visits Fab Phones and is shown three different phones.

The details are shown below:

Which phone should Sarah choose? Why?

Sinsang R2
£26 per month
Unlimited Texts
500 MB Data

Topple Light 450
£31 per month
Unlimited Texts

Pear Phone Blue
£21·50 per month
Unlimited Texts

2 The Thomson family are looking to find the best deal for upgrading their internet to broadband.

They consider these three options:

Real Broadband	_Happy Internet Company_	_BRIT Net_
£2·50 a month plus	£3·25 a month plus	£14 a month plus
£14 line rental	£14·95 line rental	£10·75 line rental
Monthly Usage Limit: 10 GB	Monthly Usage Limit: 20 GB	Monthly Usage Limit: 60 GB

Which option should the Thomson family choose and why?

3 The Thomson family also wants to change their telephone provider.

They are offered three different deals: £12·50 per month plus £0·15 per minute for calls, £14·95 per month plus £0·12 per minute for calls, £34·25 per month including calls. The Thomson family estimate that they will make 2 hours 30 minutes of calls a month. Which deal should they choose? Give a reason for your choice.

4 Hannah and Rachel want to go on holiday in Europe. They have a limited budget and have to choose where to go based on cost rather than location. Here are three choices:

City Break to Paris, France, for 3 nights costing £189 per person

Beach Holiday to Costa del Sol, Spain, for 7 nights costing £294 per person

Camping in Lake Garda, Italy, for 8 nights costing £392 per person.

a Which holiday is the cheapest?

b Does this holiday give the best value for money?

c Where should Hannah and Rachel go on holiday? Give a reason for your choice.

Hannah has £250 spending money.

A local bank is offering an exchange rate of €1·20 to the pound (£).

Hannah could change her money once she arrives at the airport at a rate of €1·23 to the pound (£) with 1·5% commission charged before the money is changed.

d Where should she change her money and how many euros will she have to spend?

5 A pharmacy normally sells a 200 ml bottle of shampoo for £4·85.

The pharmacy decides to run two special offers:

BIGGER BOTTLE 50% extra shampoo for £7·28	Price cut Price reduction of 35% on 200 ml bottle

Which offer is the better deal? Justify your answer.

Preparation for assessment

1 Tabitha travelled from Glasgow to Florida. She changed £800 into dollars at a rate of $1·62 to the £1.

How many dollars did she receive?

2 Harry is returning to Edinburgh from a trip to Australia.
He changes 120 Australian dollars into pounds sterling at a rate of 1·52 Australian dollars to the £1.
How many pounds sterling does Harry get?

3 Georgia came over from San Francisco to visit Europe. She intended to stop first in London.
She changed $3000 at a rate of £0·62 to the $1.
   a How many pounds sterling did she get?
   b Having spent £750, she converted the remainder to euros, getting €1·22 to the pound.
      How many euros did she get?
   c She spent €675 and converted the rest to dollars to go home.
      The rate quoted was €0·75 per dollar. How many dollars did she get?

4 Calculate the simple interest earned on:
   a £50 at 6% p.a. for 12 months
   b £400 at 3% p.a. for 6 months
   c £7000 at 1·5% p.a. for 3 months
   d £80 000 at 2·6% p.a. for 7 months.

5 Martha and Joy invest £1600 and £1400 respectively in a Super Saver account with a rate of interest of 1·82% p.a. compound interest.
   a If Martha withdraws her money after three years, how much interest will she earn?
   b Joy takes her money out after four years.
      Who earned more interest and by how much?

6 The table shows the monthly repayments on a loan of £1000.

APR (%)	Loan term (months)				
	12	18	24	36	48
4	£85·15	£57·33	£43·42	£29·52	£22·58
5	£85·61	£57·78	£43·87	£29·97	£23·03
6	£86·07	£58·23	£44·32	£30·42	£23·49
7	£86·53	£58·68	£44·77	£30·88	£23·95

For each of the following loans, calculate:

    i the monthly repayment   ii the total repayment   iii the cost of the loan.

a £1000 borrowed for 12 months with APR of 7%.

b £9000 borrowed for 48 months with APR of 4%.

c £7600 borrowed for 24 months with APR of 5%.

7 Mr Banks is going to get a loan for £12 000 from his building society. He is offered two different loans.

Option	1	2
Loan term (months)	12	3
APR (%)	10·2	8·4

a Calculate:

    i the total amount due

    ii the monthly payments for each option.

b Which loan should Mr Banks take out? Justify your answer.

8 Louise gets a new credit card with a monthly rate of 1·35% (APR 17·5%).

a Calculate the interest Louise would owe at the end of the first month if she spent £10 500 on her credit card.

b How much interest would Louise have to pay in a year if she took no more out on the card and she never paid any of the debt off?

9 Christine is looking to buy some rewritable DVDs. The following options are available:

      5-pack for £5·99

      10-pack for £8·87

      25-pack for £15·01.

a Work out which pack is the best buy.

b Why might Christine not choose the best buy?

10 Remember Chris? He still hasn't decided whether to use his credit card or get a bank loan to help him purchase a car.

He wants to borrow £2500.

Which choice would give Chris the better deal?
Justify your answer.

Blue Sky Credit Card

APR 15·9%

BANK LOANS

Borrow £1000
12 monthly payments of £93·02

Before we start...

Tommy is preparing his vegetables for the annual horticultural show.

He is deciding on which of his potatoes to enter.

He needs a group of five 'tatties'.

They must be in good condition, and while it is important that they are as big as possible, it is more important that they are all of a similar shape and size.

After cleaning his potatoes he selects three groups of five so that the 'tatties' in each group have a very similar shape.

He then weighs the potatoes.
Here are the weights in grams:

Group 1:   395   406   425   428   442

Group 2:   411   427   435   448   469

Group 3:   412   412   431   443   459

How would you advise Tommy to choose the group of potatoes that has the best chance of winning?

What you need to know

1   Calculate:

  a 32 + 40 ÷ 2   **b** (32 + 40) ÷ 2   **c** 48 + 51 + 63 ÷ 3   **d** (48 + 51 + 63) ÷ 3

2   Find the total of each group of numbers.

  a 3, 5, 2, 4, 8   **b** −15, 18, 12, −11, 15, 15   **c** 4, 2, −1, −2, −5, 3, 1

3   For each list, calculate the difference between the highest and lowest number.

  a 23, 48, 32, 14, 52, 49, 37   **b** 4·1, 3·6, 6·9, 2·5, 4·7, 3·2

  c 7, −3, −7, 4, 6, 2, −1, 0, 3

4   Select the middle number in each list.

  a 3, 4, 6, 8, 11, 12, 14, 16, 19   **b** 126, 129, 135, 138, 140

  c 3·4, 4·1, 5·7, 5·8, 6·1, 6·1, 6·3

5.1 The mean

You may often hear someone, in general conversation, talking about an 'average' price or an 'average' time or an 'average' size.

When they do this they are talking about a price, a time, a size that is fairly typical or 'in the middle'.

'Average' can be a term that people use loosely without actually doing an accurate calculation.

There are three different commonly used averages: the **mean**, the **median** and the **mode**.

The **mean** is calculated by adding together all the items in a data group and then dividing this sum by the number of items in the data group.

Example 1

During the month of April a golfer has the following scores:

    83   84   79   77   81   77   75   75

He wants to know his mean score for this month.

Calculate the mean of his scores.

Total of scores = 83 + 84 + 79 + 77 + 81 + 77 + 75 + 75 = 631.

He had 8 rounds of golf during April.

$$\Rightarrow \text{Mean} = \frac{631}{8} = 78 \cdot 875.$$

So his mean score for the month was 78·9 (to 1 d.p.).

Example 2

A farmer goes to the 'tup sales' (ram sales) and buys eight rams.

They were all different prices but he spends £9370 in total.

What is the mean price of a ram?

$$\text{Mean} = \frac{£9370}{8} = £1171 \cdot 25.$$

So the mean price of a ram is £1171·25.

Exercise 5.1A

1   Calculate the mean of each set of numbers.

    a   3, 6, 8, 4, 4, 5

    b   13, 15, 14, 17, 20, 10, 9

    c   100, 102, 109, 110, 105, 112, 108, 110

    d   4·1, 5·2, 3·6, 3·4, 4·5, 3·2, 3·1, 3·1, 3·1

    e   625, 627, 642, 655, 623, 528, 656, 671, 645, 648

2   When John was looking for a new television he shopped around.

He found the same type for sale in five different places.

The prices were:

£500   £470   £450   £525   £520.

Calculate the mean price.

3   Isobelle was going to a wedding and needed something to wear.

She found a dogtooth print dress that she really liked.

She found the same dress on seven different websites.

The prices were:

£79   £80   £75   £80   £85   £79   £80.

Calculate the mean price of the dress.

4   The results in the women's long jump in a Diamond League athletics meet in Zurich in August 2012 were:

6·92 m, 6·85 m, 6·80 m, 6·79 m, 6·54 m, 6·53 m, 6·50 m, 6·45 m, 6·31 m and 6·10 m.

Calculate the mean distance.

5   Elaine bought a pack of courgettes from the supermarket.

The weight of the pack was 400 grams.

The weight of the packaging was negligible.

There are three courgettes in a pack.

What is the mean weight of a courgette?

6   Sally's monthly gas bills last year were:

£52·49 £57·32 £64·83 £71·90 £79·53 £58·94 £87·61 £81·67 £76·07 £68·55 £62·72 £51·12.

She didn't like having big bills in the winter, so she decided to take advantage of an easy-payment scheme.

Her payments are adjusted so that every month she pays the same.
What she pays is the mean monthly bill for last year. If she uses more, she'll be billed for the extra at the end of the year.

Calculate her new monthly payment.

Example 3

In a cycling hill climb time trial the times of six competitors were:

19:51 (19 minutes, 51 seconds), 21:56, 22:23, 23:02, 24:39 and 25:21.

What is the mean time expressed in minutes and seconds?

The times are given in minutes and seconds.

We can express all of the times in seconds.

$$19{:}51 = 19 \times 60 + 51 = 1191$$
$$21{:}56 = 21 \times 60 + 56 = 1316$$
$$22{:}23 = 22 \times 60 + 23 = 1343$$
$$23{:}02 = 23 \times 60 + 2 = 1382$$
$$24{:}39 = 24 \times 60 + 39 = 1479$$
$$25{:}21 = 25 \times 60 + 21 = 1521$$

Total = 1191 + 1316 + 1343 + 1382 + 1479 + 1521 = 8232.

Mean $= \dfrac{8232}{6} = 1372$ seconds.

We now need to express this in minutes and seconds.

$\dfrac{1372}{60} = 22$, remainder 52.

So the mean time is 22:52, or 22 minutes 52 seconds.

Example 4

Geoff is putting together a four-man team for a tug-of-war competition.

The rules state that the mean weight of a team must not exceed 110 kilograms.

He already has three team members and is looking for a fourth.

Geoff himself weighs 115 kg, his other two team members, Jim and Johnny, weigh 112 kg and 118 kg respectively.

What is the heaviest that the fourth member of Geoff's team can be?

Total weight of team = mean weight $\times$ 4 = 110 $\times$ 4 = 440 kg.

Total weight of the existing three team members = 115 + 112 + 118 = 345 kg.

Maximum weight for fourth team member = 440 − 345 = 95 kg.

So the fourth member of the team must not weigh more than 95 kg.

Exercise 5.1B

1 In a cross-country team race there are five members in each team.

The mean time for the quickest **four** runners in each team is calculated.

The winning team is the one with the lowest mean time.

Here are the times for the top three teams.

Surefoot Runners: 25:34 (25 minutes, 34 seconds), 27:37, 28:14, 26:52, 26:41.

Hartree Harriers: 27:49, 25:24, 26:53, 31:38, 26:44.

Westerfield AC: 26:32, 26:47, 25:53, 26:48, 26:35.

a Calculate the mean times of the quickest four runners from each team.

b Which team won?

2 George and two friends are entering the school quiz.

Quiz teams must be of three, four or five players but their mean age must be 14 years or less.

a George is 14 and his friends are both 15. Can they enter the quiz?

b They decide to ask a younger friend to join them to make a team of four.
What is the oldest that this friend can be so that the team is eligible to enter?

c This friend says no. So they ask two other younger friends (both the same age) to join the team.
What is the oldest that these friends can be?

3 This table shows the temperature in degrees Celsius at 8 a.m. on each day for a week in January.

Sunday	Monday	Tuesday	Wednesday
4	1	0	1

Thursday	Friday	Saturday
−1	−2	−2

What is the mean temperature? Give your answer to 1 decimal place.

4 Jack's class were sitting a test but Jack was absent.

The other 25 students in the class scored a mean mark of 49%.

Jack sat the test later and scored 75%.

What was the new mean mark for the class?

5 Eilidh is a salesperson.

If her 'average' (mean) monthly sales for August, September, October and November is more than £25 000, then she will get a Christmas bonus.

Her sales for August, September and October were £27 000, £24 000 and £23 000.

What does she need her sales to be in November if she is to get her Christmas bonus?

5.2 Making comparisons using the mean

Is one set of figures better than another?

Sometimes it's obvious but often it is very difficult to compare two groups of data.

One way of comparing two groups of data is by comparing the means.

It may be useful to find the difference between the means ... and more useful to give the difference as a percentage of the original mean.

Example 1

Holly has always been keen to do well at school.

At the end of third year she scored the following marks:

English 63, Maths 87, Chemistry 68, Physics 82, French 65, History 59, Geography 62, Technical Studies 76.

She made up her mind to try to improve her results in her November exams in S4.

Here are her marks:

English 65, Maths 88, Chemistry 64, Physics 81, French 71, History 66, Geography 64, Technical Studies 72.

All exams were out of 100.

Has she made an overall improvement?

End of S3:

$$63 + 87 + 68 + 82 + 65 + 59 + 62 + 76 = 562$$

$$\frac{562}{8} = 70 \cdot 25.$$

So Holly's mean mark at the end of S3 was 70·3 marks per exam (to 1 d.p.).

November of S4:

$$65 + 88 + 64 + 81 + 71 + 66 + 64 + 72 = 571$$

$$\frac{571}{8} = 71 \cdot 375.$$

So Holly's mean mark in November S4 was 71·4 marks per exam (to 1 d.p.).

Her mean mark has improved from 70·3 to 71·4.

This represents a difference of 1·1 marks per exam.

This represents an improvement of $\frac{1 \cdot 1}{70 \cdot 3} \times 100 = 1 \cdot 6\%$ (to 1 d.p.).

So she has made a small overall improvement in her results between the end of S3 and November S4.

Example 2

During a survey on the flow of traffic in the town centre, people were asked to grade how satisfied they were with the situation on a scale of 1 to 10. The mean grade was 6·4.

A month later, after changes, another group of people were asked. This time the mean was 7·5.

Comment on the change in satisfaction.

The difference is $7 \cdot 5 - 6 \cdot 4 = 1 \cdot 1$ ... an improvement in the score of 1·1. (This is the same as in Example 1.)

The percentage improvement $= \frac{1 \cdot 1}{6 \cdot 4} \times 100 = 17 \cdot 2\%$.

So the improvement in the score of 1·1 is a relatively large increase.

Exercise 5.2A

1 For each pair of lists,
   i calculate the means
   ii identify which list has the higher mean
   iii express the difference as a percentage of the mean for List A.

   a List A: 16, 18, 7, 12, 13, 7, 15, 14, 8, 7, 14, 9.
   List B: 12, 15, 8, 11, 9, 14, 10, 15.

   b List A: 73, 86, 76, 82, 79.
   List B: 65, 75, 73, 86, 93, 96, 72, 76, 68, 75.

   c List A: 235, 418, 472, 394, 274, 306, 385, 297.
   List B: 379, 358, 381.

   d List A: 6·2, 4·7, 15·5, 9·3, 12·8, 7·5, 8·3, 11·7, 9·3.
   List B: 10·5, 11·3, 8·4, 9·8, 12·7, 10·1.

2 In a knitwear factory there were concerns about the number of garments that had faults.

It was decided to keep a record of how many faulty garments were produced by each hand knitter.

Over the next week the eight hand knitters were monitored for the number of faulty garments they produced. Here are the results:

   5   4   7   5   1   6   7   6

   a Calculate the mean number of faulty garments produced per knitter.

   b A week later, new working practices were introduced.

   These were intended to reduce the number of faulty garments.

   Here are the results:

   5   5   5   4   1   4   6   3

   Calculate the mean number of faulty garments produced per knitter in this week.

   c Has the change to working practices brought about a reduction in the number of faulty garments?

   d Is there anything else that should be taken into account before the experiment can be declared a success?

3 Mrs Currie kept 12 hens. In one week she kept a record of how many eggs they laid.

   6   6   5   7   4   5   3   7   7   5   6   5

   a Calculate the mean number of eggs laid per hen in this week.

   b The following week she started using a new feed.

   Unfortunately in the same week, a fox killed one of the hens.

   Here is the record of how many eggs were laid.

   6   6   6   7   6   5   4   6   6   6   6

   What is the mean number of eggs laid per hen in the second week?

   c Do you think that the change of feed has improved the laying record of the hens?

   Consider the percentage change in your response.

4   On Saturday 19th January 2013 in the Scottish Premier League the goals scored by the ten teams who played were:

4, 1, 2, 3, 1, 4, 3, 0, 1, 1.

a   Calculate the mean number of goals scored per game in the Scottish Premier League.

b   In the English Premier League on the same day there were 14 matches and the goals scored by the teams were:

5, 0, 2, 0, 1, 2, 3, 1, 1, 1, 2, 3, 2, 2.

Calculate the mean number of goals scored per game in the English Premier League.

c   In which league was the scoring rate higher?

5   Andrew runs a fencing firm.

He buys posts from two timber companies.

When erecting the fences some of the posts break because of poor quality of timber.

Andrew keeps a record of the percentages of broken posts in each load of timber over a six-month period.

He bought nine loads from John James & Sons and the percentages of broken posts were:

4%   5%   4%   3%   5%   4%   1%   4%   4%.

He bought five loads from The Robert Johnson Timber Company and the percentages of broken posts were:

3%   4%   3%   4%   3%.

Andrew has decided to use The Robert Johnson Timber Company as his main supplier.

By comparing the means of these two groups of data can you say if you think he is making the correct decision. Give a reason for your answer.

Example 3

Alan is a very keen sprinter.

Here are his times (in seconds) for the major 100 m races that he ran as a 14-year-old.

12·9, 12·9, 12·8, 12·7, 12·8, 12·7, 12·6, 12·5, 12·4, 12·4, 12·5, 12·4, 12·5, 12·4, 12·4.

Here are his times in the major 100 m races that he ran as a 16-year-old.

12·1, 12·1, 12·0, 12·0, 11·9, 11·8, 11·8, 11·8, 11·8, 11·8, 11·7, 11·7, 11·9.

Calculate his mean time as a 14-year-old and his mean time as a 16-year-old and calculate the improvement as a percentage of the original mean.

Mean time as a 14-year-old:

$12·9 + 12·9 + 12·8 + 12·7 + 12·8 + 12·7 + 12·6 + 12·5 + 12·4 + 12·4 + 12·5 + 12·4 + 12·5 + 12·4 + 12·4 = 188·9.$

The mean time $= \dfrac{188·9}{15} = 12·5933 \ldots$

$= 12·59$ seconds (to 2 d.p.).

Mean time as a 16-year-old:

12·1 + 12·1 + 12·0 + 12·0 + 11·9 + 11·8 + 11·8 + 11·8 + 11·8 + 11·8 + 11·7 + 11·7 + 11·9 = 154·4.

The new mean time $= \dfrac{154.4}{13} = 11.8769...$

$= 11.88$ seconds (to 2 d.p.).

Improvement $= 12.59 - 11.88 = 0.71$ second.

Percentage improvement $= \dfrac{0.71}{12.59} \times 100 = 5.6\%$ (to 1 d.p.).

Example 4

Fergus Alexson is the manager of Redvest United football club. He regularly complained about the small amounts of stoppage time at the end of matches in which his side lost. He claimed that referees didn't add on enough time when he was getting beaten.

A journalist decided to see if there was any truth in his claim. He recorded a stoppage time of 4 minutes 42 seconds when the Redvests lost as 4:42 L.

In the last ten matches played by the club the stoppage times were:

4:42 L, 3:49 W, 3:55 W, 4:17 W, 5:31 L, 3:37 W, 3:47 W, 4:56 L, 4:13 W, 3:52 D.

a Calculate the mean stoppage time in matches in which the team:
  i won   **ii** lost   **iii** drew.

b **i** Compare the mean stoppage time in matches won with the mean stoppage time in matches lost.
  ii Compare the mean stoppage time in matches lost with the mean overall stoppage time. Comment on Fergus' claim.

a **i** Change the stoppage times of games won from minutes and seconds into seconds.

Games won: 229 seconds, 235 seconds, 257 seconds, 217 seconds, 227 seconds, 253 seconds

Total time = 229 + 235 + 257 + 217 + 227 + 253 = 1418 seconds.

Mean stoppage time in games won $= \dfrac{1418}{6} = 236.3$ seconds.

ii Games lost: 282 seconds, 331 seconds, 296 seconds.

Total time = 282 + 331 + 296 = 909 seconds.

Mean stoppage time in games lost = $\frac{909}{3}$ = 303 seconds.

iii Games drawn: total time = 232 seconds.

Mean stoppage time in games drawn = $\frac{232}{1}$ = 232 seconds.

b **i** Difference in stoppage times: in games they lost they got 66·7 seconds more stoppage time (on average) than in games they won.

ii Total of all 10 games = 1418 + 909 + 232 = 2559 seconds.

Mean stoppage time overall = $\frac{2559}{10}$ = 255·9 seconds.

The mean stoppage time in matches lost is 47·1 seconds more than this.

As a percentage of the overall mean: $\frac{47·1}{255·9}$ × 100 = 18% (to the nearest whole number).

When they are losing, the Redvests get 18% more stoppage time than usual.

Fergus has no cause for complaint.

Exercise 5.2B

1 Alistair is looking for a new car.

He sees a five-year-old Ford Focus at his local showroom that has done 40 000 miles. It is offered at a price of £4795.

He decides to do a bit of research before going into the showroom.

He looks on the internet and finds 10 five-year-old Ford Focuses with a similar mileage.

Here are the prices:

£4699   £4549   £4895   £4999   £4799   £4595   £4649   £4495   £4645   £4595

a Calculate the mean of these prices.

b Is the car in the local showroom above or below the mean of these internet prices?

c Alistair decides to buy the car from his local showroom. Can you explain why?

2 David is a dairy farmer.

His cows are kept inside during the winter and go out to graze in the fields in May.

Every year, when they are let out, David expects to see a rise in the milk produced (the milk yield).

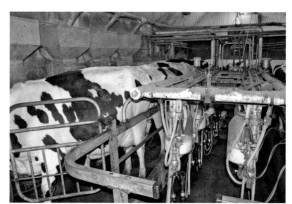

a Here are the milk yields, in litres per day, of 12 cows in the week before they were let out into the fields:

24·3   12·4   25·8   33·9   24·7   24·3
33·7   23·2   26·7   27·4   11·3   22·8.

Calculate the mean milk yield for these cows at this time.

b Here are the milk yields for the same 12 cows after they have been let out to the fields:

30·4   38·6   31·7   29·3   40·7   30·4   18·8   19·1   32·6   33·1   26·9   27·3.

Calculate the mean milk yield for these cows now that they have been let outside to graze.

c Would you say there has been an increase in the milk yields of the 12 cows?

3 Norrie always played golf once a week.

His friend advised him that if he played more often then his game would improve.

He decided to take this advice to see if it helped. He looked back over his scores of the last eight weeks.

Date	2 June	9 June	16 June	23 June	30 June	7 July	14 July	21 July
Score	85	82	83	86	85	83	79	85

a Calculate his mean score for these rounds.

b Over the next eight weeks he played twice a week and kept his scores.

Date	28 July	29 July	4 Aug	5 Aug	11 Aug	12 Aug	18 Aug	19 Aug	25 Aug	26 Aug	1 Sep	2 Sep	8 Sep	9 Sep	15 Sep	16 Sep
Score	84	92	83	90	82	87	80	83	79	82	79	81	78	79	79	77

   i Calculate the mean score for these rounds of golf.
   ii By how much did his mean score improve?

c Norrie tells his friend of this improvement.

His friend says, 'I thought that you'd improved far more than that. I think you should look more closely at your scores since you started playing twice a week. Split them into two sets of eight rounds. Find the mean for each of the sets.'

Find Norrie's mean score for:
   i the first set of eight rounds
   ii the second set of eight rounds.

d Comment on any improvement.

4 Steve and Christina were looking at their phone bill.

There was a list of calls that lasted longer than 10 minutes.

The lengths of these phone calls were:

13:15 (13 minutes, 15 seconds), 11:03, 11:55, 15:35, 10:17, 55:27, 46:38, 23:42, 41:03.

a Change each time into seconds.

b Calculate the mean length of a call that lasts longer than 10 minutes.

c They decided that they would try to reduce the **number** of calls that lasted longer than 10 minutes and also the **mean length** of calls over 10 minutes.

Here is the list of calls over 10 minutes in their next bill:

14:38, 28:23, 18:27, 11:17, 11:55, 10:03, 19:42, 13:42, 12:28, 10:35, 17:05.

   i Change the times into seconds.
   ii Calculate the mean length of a call that lasts longer than 10 minutes.
   iii Did they reduce the number of these calls?
   iv Did they reduce the mean length of these calls?

5.3 The range

It is helpful to get some indication of how spread out a data group is.

The range is one such measure of the 'spread' of data.

It is defined as the difference between the greatest and least piece of data.

Range = highest − lowest.

Example 1

A travel agent was doing a promotion on 'Villa with a Pool' holidays.

He listed five prices: Menorca £548, Mallorca £463, Kefalonia £590, The Algarve £505, Zante £675.

Summarise the data by calculating:

a the mean price

b the range of prices.

a Total = £548 + £463 + £590 + £505 + £675 = £2781.

Mean = $\dfrac{£2781}{5}$ = £556·20.

b Highest price is £675. Lowest price is £463.

Range of prices = highest − lowest
= £675 − £463
= £212

The range of prices is £212.

Exercise 5.3A

1 Find the range of each data set.

a 5, 2, 6, 7, 3, 9, 3, 5, 6, 3, 7, 4, 6

b 54, 37, 28, 64, 73, 88, 24, 16, 56, 63

c 94, 121, 104, 113, 97, 125, 116, 119

d 2300, 3200, 4700, 3600, 9200, 2600

e 341, 426, 629, 544, 586, 382, 528

f 56, 3, 4593, 385, 452, 94, 77633

2 In August 2012, Dundee FC returned to the Scottish Premier League.

Here are the attendances for their first 10 home games at Dens Park.

5984   4905   5176   6158   5006   5344   6743   11 419   9276   9013

Summarise the data by finding the mean and range.

3 Over the 10 days from 23rd January to 1st February 2013 the wind speed in Fochabers was recorded in miles per hour:

7   11   7   11   6   12   17   16   16   12.

Find the range of these wind speeds.

4   A food critic has a column in a weekend newspaper.

Each week he visits a restaurant, has a meal and then writes about it for the newspaper.
He also gives each restaurant a mark out of 30.

Here are the marks that he gave to 12 restaurants:

26   25   24   17   22   23   19   23
23   25   18   19.

Summarise the marks by finding the mean and range.

Exercise 5.3B

1   Find the range of these data groups.

a   23·5, 29·4, 26·8, 21·5, 26·7, 24·6, 27·4

b   0·8, 0·3, 1·4, 0·7, 0·4, 1·2, 0·6, 1·1

c   12, 7, 4, −3, 8, 7, −1, 4, 11, 9

d   25, 13, −5, −17, 4, 9, −1, 21

e   8, 3, 1, 7, −4, 4, −2, −1, 1, 3

f   100, 74, 2, −5, 9, 85, −25

2   Alesha is self-employed. Over the last eight years she has made a profit on six occasions and a loss on two occasions.

Here are the figures from the last eight years.

Year ending 5th April ...	2005	2006	2007	2008	2009	2010	2011	2012
Profit or loss	£10 000 profit	£15 000 profit	£18 000 profit	£22 000 profit	£15 000 profit	£1000 loss	£2000 loss	£7000 profit

By treating a loss as a negative profit, e.g. a loss of £2000 becomes a profit of −£2000, find the range of these profits.

3   The UK's Gross Domestic Product (GDP) is a measure of the size of the country's economy. The change in the GDP (sometimes called the growth rate) is recorded every quarter (three months) and the results are used to make judgements by economists and politicians. From the first quarter of 2008 to the last quarter of 2012 the changes in GDP were:

0·1%, −1·8%, −2·1%, −1·5%, −0·2%, 0·4%, 0·4%, 0·6%, 0·7%, 0·6%, −0·4%, 0·4%, 0·1%, 0·6%, −0·3%, −0·2%, −0·4%, 0·9%, −0·3%, 0·2%.

Calculate the range of these figures.

4   The growth rate for Scotland's economy from the summer of 2008 until the end of 2012 was:

−1·8%, −1·2%, −0·8%, −1·1%, −0·6%, 0·3%, 1·2%, 0·5%, 0·5%, −0·3%, 0·1%, −0·1%, 0·2%, 0·4%, −0·5%, −0·1%, 0·6%.

Calculate the range of Scotland's growth rate over this period.

5.4 Comparing data groups using the range (and mean)

Example 1

On 24th January 2013 it snowed in Innerleithen.
On 25th and 26th January the temperature in °C
was recorded every three hours.

Saturday 25th January								Sunday 26th January							
00 00	03 00	06 00	09 00	12 00	15 00	18 00	21 00	00 00	03 00	06 00	09 00	12 00	15 00	18 00	21 00
0	0	−1	−1	0	0	0	0	1	2	1	2	2	2	2	2

Compare Saturday's with Sunday's temperature.

Saturday

$$\text{Mean} = \frac{\text{sum}}{8} = \frac{-2}{8} = -\frac{1}{4}\,°C.$$

Range $= 0 - (-1) = 1\,°C.$

Sunday

$$\text{Mean} = \frac{\text{sum}}{8} = \frac{14}{8} = 1\tfrac{3}{4}\,°C.$$

Range $= 2 - 1 = 1\,°C.$

Comparing the means: Sunday's mean is a lot better than Saturday's.

Comparing the ranges: The ranges are the same. Neither of the data sets is very variable.

Example 2

Miss Martin teaches a class of 24 P6 pupils in a city primary school.

Here are the heights of her pupils in centimetres:

145	141	145	137	147	146	142	143	139	141	140	144
140	142	140	149	131	142	148	147	142	142	143	141

Her friend, Miss Greig, teaches in a small country primary school.

Her class has 20 pupils and is a composite class made up of pupils from P5, P6 and P7.

Their mean height is 142·3 cm with a range of heights of 29 cm.

Compare the heights of the two classes.

Total of heights of Miss Martin's pupils = 3417 cm.

So the mean height of Miss Martin's class = $\dfrac{3417}{24}$ = 142·4 cm (to 1 d.p.).

Range of heights of Miss Martin's class = 149 − 131 = 18 cm.

Conclusion

Although the classes have means which are very similar there is a big difference in their ranges.

This means that on average the pupils were of comparable height but there is much more variation in the heights of the pupils in Miss Greig's class.

Exercise 5.4A

1 Compare each pair of data sets by first finding means and ranges.

   a A: 6, 9, 2, 4, 4, 4, 4, 6, 3, 7, 2   B: 6, 6, 6, 7, 3, 5, 5, 6

   b C: 6, 3, 7, 11, 7, 4, 13, 12   D: 15, 17, 23, 16, 22, 22, 17, 18, 22, 21, 19

   c E: 41, 47, 64, 73, 37, 68, 54, 59   F: 52, 46, 70, 27, 63, 75, 24, 50, 41

2 Before his exams, Cameron practises by doing past papers and marking them using the marking instructions, which are on the SQA website.

   a He does five maths past papers, scoring 64%, 73%, 79%, 71% and 81%.
   Calculate the mean and range of his maths scores.

   b He does five chemistry past papers, scoring 58%, 56%, 62%, 64%, 63%.
   Calculate the mean and range of his chemistry scores.

   c Compare his two performances.

   d Which score do you think is more predictable, his score in maths or chemistry?

3 A doctor compares the weights, in kg, of two groups of people in different areas of the city.

   a The weights of the Silverhills residents are 75, 84, 78, 86, 72, 81, 89, 68, 75, 77.
   Calculate the mean and range of these weights.

   b The residents in the Garnetbraes area have a mean weight of 86·4 kg and a range of 36 kg.
   On average, which set of residents is heavier?

   c Write two sentences comparing the two sets of residents.

4 A biology class investigates populations of earthworms.

   They go into a field where the soil is fairly uniform. Each student marks off a square metre and then digs to see how many earthworms they can find in the top 10 centimetres of soil.

   Here are their results:

        5 6 5 6 4 3 7 4 5 6 6 4 5 6.

   a Calculate the mean and range of these numbers of earthworms.

   b They repeat the experiment in a field that has some dry areas and some wet areas.
   They get these results:

        4 5 3 6 5 5 5 4 2 5 9 2 8 1.

   Compare the two sets of results.

Exercise 5.4B

1 A discount book company sells books to the teachers in a school.

In one month there were orders from 11 teachers.

The prices of these orders were:

£14, £5·70, £8·20, £21·50, £12, £3·50, £6·90, £5·40, £4·20, £9·40, £8·50.

a Calculate the mean and range of these prices.

b There is also a delivery charge of £1 with each order.
Add £1 to the price of each order and calculate the mean and range when the delivery charge is included in the price.

c What difference has this made to the mean and range?

2 Emma looks at the balance of her current account at the end of each month for the tax year to 5th April 2010.

Month	Apr	May	Jun	Jul	Aug	Sep	Oct	Nov	Dec	Jan	Feb	Mar
Balance	£158	£163	£175	£180	−£30	£10	£63	£75	−£214	−£147	−£72	−£27

a Calculate the mean and the range of these balances.

Emma is amazed that there is so much variation in the monthly balance.

She is also disappointed that her account is 'in the red' (overdrawn) at the end of the tax year.

In any month when it looks as if she will have more than £100 she will put the difference into a savings account. For expensive months she can get the money from her savings account. This should allow her to avoid going overdrawn when she has an expensive month.

Here are the balances for Emma's current account at the end of each month for the tax year to 5th April 2011.

Month	Apr	May	Jun	Jul	Aug	Sep	Oct	Nov	Dec	Jan	Feb	Mar
Balance	£33	£73	£100	£100	−£50	£23	£97	£100	−£55	−£42	£47	£100

b Compare the range of 2010 with 2011.
Has she reduced the variation in her end-of-month balance?

c Compare the mean of 2010 with 2011.
On average does she have more or less in this account at the end of the month?

3 Julie has been offered a salesperson's job with two different companies, Super Sales and Perfect Pitch.

In both jobs, commission is added to the monthly salary.

So that she can make a more informed decision on which job to accept, she asks to see the monthly payments for a salesperson in each company over the last year.

Super Sales (£)	1800	2000	2100	1950	2150	1825	1200	2050	2150	1955	2150	2000
Perfect Pitch (£)	2005	2010	1500	1400	2500	2850	2750	1995	1600	1550	2010	2015

a   Which job do you think she should choose ?

b   She noticed that one of the payments made by Super Sales (£1200) was very much lower than the other 11 payments. She was told that the salesperson had been absent through illness in this month.

   She then decides to calculate the mean and range for the 11 other payments.

   Calculate the mean and range of these 11 payments.

c   Might this affect her decision?

4   Sharon is a busy person.

She finds doing her shopping on the internet to be a big time saver.

However she has to be at home when her purchases are delivered. This means arranging time off work.

If the company deliver at the time that they say they will, then she can be back in work an hour after leaving.

If the delivery is early, then she'll miss it. If the delivery is late, then she will be late back to work.

While on holiday she noted the performance of two companies. The figures are in minutes and negative times are early.

Swallows	−40	−45	−95	−175	−185	50	55	89	190	190
Aztecs Direct	−15	−10	−7	12	20	25	27	30		

a   i   Calculate the mean delivery time for Swallows.
    ii Calculate the mean delivery time for Aztecs Direct.

b   According to the means, which company, on average, appears to deliver closest to the agreed delivery time?

c   If they come early when Sharon is at work, she will miss them.

   How might this affect her decision?

5.5 Other averages – median and mode

The mean is only useful when the data is numeric.

However, sometimes it does not produce what you would call a typical measure of what is happening.

Consider two firms that both have five people working for them.

Their wage rates in pounds per hour are tabulated:

Person	1	2	3	4	5
Firm A	10	10	10	10	200
Firm B	15	15	15	15	15

The mean rate for Firm A is £48 per hour
the mean rate for Firm B is £15 per hour
... but which firm would you prefer to join?

Two other measures of the 'centre' can be defined.

The median

The **median** is the middle score when the data set is listed in order of size.

If there is an even number of entries in the data group, then the median is the mean of the middle two numbers.

If there are n numbers in the ordered list, then the median is the $\frac{n+1}{2}$ th position.

The mode

The **mode** is the score that occurs the most often.

It is the only one of the averages that can be used with qualitative data.

Example 1

A snack shop was studying the number of cheese and pickle sandwiches that they sold on weekdays over a three-week period:

52, 45, 47, 48, 62, 54, 35, 41, 47, 59, 49, 48, 52, 52, 62.

a Find the median number of cheese and pickle sandwiches sold.

b What is the modal number of cheese and pickle sandwiches sold? (Find the mode.)

a Arrange the numbers in order.

(This can be done using a stem-and-leaf diagram if the data set is large.)

35, 41, 45, 47, 47, 48, 48, 49, 52, 52, 52, 54, 59, 62, 62

There are fifteen numbers in the list.

The median occupies position $\frac{15+1}{2}$ = 8th position.

35, 41, 45, 47, 47, 48, 48, 49, 52, 52, 52, 54, 59, 62, 62

So the median is 49 cheese and pickle sandwiches.

b Putting the numbers in order also helps us to find the mode.

We can see that there are two 47s, two 48s and two 62s, but there are three 52s.

So the modal number of cheese and pickle sandwiches sold is 52.

Exercise 5.5A

1 State the median and mode of each list.

a 5, 6, 7, 9, 11, 11, 12
b 32, 35, 35, 36, 39, 39, 39, 42, 45
c 3·1, 4·4, 4·5, 4·7, 5·0, 5·1, 5·1
d 6, 2, 3, 2, 7, 11, 6, 8, 3, 7, 2   (Careful!)

2 In a biology experiment, 21 different packets of 12 seeds were laid on damp gauze.

After three days the number of the seeds in each batch which have germinated is counted.

4, 7, 5, 4, 6, 8, 5, 9, 3, 6, 10, 5, 6, 5, 7, 4, 6, 5, 7, 4, 5

a What is the median number of germinated seeds?

b What is the mode?

3 Employees who work in an office were asked how long it took them to travel to work.

Here are their journey times in minutes.

25, 27, 16, 13, 34, 39, 22, 29, 31, 52, 7, 31

a What is:
   i the median journey time
   ii the modal journey time?

b Suppose the highest time is ignored.

Which of the two averages is affected?

4 Here is a list of the sizes of shoes sold by a ladies' shoe shop in one afternoon.

3, 7, 5, 9, 8, 8, 6, 8, 7, 6, 5, 10, 7, 8, 8, 8, 5, 5, 6, 4, 5

a What is the median?

b What is the modal shoe size?

c Which of these would be most informative if the shop owner was trying to restock?

5 Faisal kept a record of how long he spent on his homework each night for fifteen nights. His times are in minutes.

26, 20, 31, 24, 33, 5, 0, 33, 22, 25, 27, 33, 31, 0, 7

a Work out:
   i the median time
   ii the modal time
   he spent on homework.

b Which of the two do you think produces a more 'typical' time?

Exercise 5.5B

1 On 30th January 2013, the points totals of the teams in the Scottish Premier League were:

52, 37, 37, 34, 33, 32, 30, 29, 28, 28, 28, 14.

Work out the median, the mode and the range of these points totals.

2 Sixteen teenagers were asked how many songs they had downloaded to their phones in the last week. The totals were as follows:

8, 12, 15, 13, 19, 21, 18, 14, 11, 15, 13, 20, 25, 23, 20, 19.

a What is the median number of songs?

b Why is it unhelpful to consider the mode?

c Another ten are questioned, answering 2, 2, 5, 6, 22, 24, 24, 24, 32, 35.
   i Which of the averages is most affected?
   ii Which of the measures of central tendency, mean, median or mode, would be most useful in this short survey?

3 Michaela is keeping a record of the species of birds which visit her bird table.

In one hour the birds that visited were:

blackbird, chaffinch, chaffinch, robin, dunnock, woodpigeon, chaffinch, blackbird, robin, chaffinch, chaffinch, blue tit, coal tit, chaffinch, jackdaw, chaffinch, blackbird, siskin.

a Name the modal bird.

b The mode is the only average that we can apply to this information. Why?

4 A millionaire has returned to his home town.

He asks 11 of his old school friends to a restaurant.

The millionaire pays for everyone's meals and drinks, spending a total of £480.

He refuses to let his friends pay for anything.

a What is the mean amount of money spent by the 12 men?

b What is the median amount of money spent by the 12 men?

c What is the modal amount of money spent by the 12 men?

d Which of the three averages is not typical of the amount spent by each person.

Preparation for assessment

1 Elaine is the practice manager at a health centre.

She is currently gathering information as part of a patient satisfaction study.

She records how long in minutes each patient was kept waiting for their appointment:

3, 2, 3, 3, 2, 4, 5, 7, 5, 3, 5, 6, 7, 5, 20, 5, 7, 3, 5, 6.

a Calculate the mean and range of these times.

b Another health centre nearby has a mean waiting time of 7·4 minutes and a range of 11 minutes.

Make two comments comparing the waiting times in the two health centres.

c State the mode and median of the data.

2 Wilson was interested in the price of houses in his street.

He found a website that gave the price of the last five houses sold in this street.

Four of them were £170 000, £215 000, £183 000, £191 000.

There was an error on the website so the fifth price couldn't be read.

a The range of the five prices was £55 000.

Calculate two possible prices for the fifth house.

b The mean of the five prices was £183 800.

What was the price of the fifth house?

3 Zoe said, 'Over the last year the mean price that I have paid for a concert ticket has been £21 and the range of prices has been £12.'

'Did you include the £1·50 booking fee for each ticket?' asked Blair. 'That's bound to affect the mean and the range.'

Is Blair right?

What will be the mean and the range once the £1·50 has been added to the price of each ticket?

4 At the start of a rugby match the players wear shirts numbered from 1 to 15.

  a What do you notice about the mean and median of these numbers?

  b What is the range of these numbers?

  c At half-time the team bring on a replacement who is wearing the number 22 shirt.

     What is the range of the shirts that are on the pitch now? (There are two possible answers.)

  d The new mean is 9·2. What is the shirt number of the player who was replaced?

5 Remember Tommy? He is preparing his vegetables for the annual horticultural show.

He is deciding on which of his potatoes to enter.

He needs a group of five 'tatties'.

They must be in good condition and while it is important that they are as big as possible it is more important that they are all of a similar shape and size.

After cleaning his potatoes he selects three groups of five so that the 'tatties' in each group have a very similar shape.

He then weighs the potatoes.
Here are the weights in grams:

    Group 1:   395   406   425   428   442
    Group 2:   411   427   435   448   469
    Group 3:   412   412   431   443   459

How would you advise Tommy to choose the group
of potatoes that has the best chance of winning?

6 Statistical diagrams

Before we start...

Angela is in charge of sales at a large company.

She is working in her office when the chief executive bursts in,

'I'm really disappointed with our sales figures.
Our sales have decreased for four years in a row.

We were aiming to be ahead of our main competitor by 2014.

I'll expect you to present these figures to the board of
directors next Wednesday, and your presentation has to be good.'

He hands her these sales figures for the last seven years.

Year	2006	2007	2008	2009	2010	2011	2012
Sales in £million	25·3	25·6	25·7	23·1	22·9	22·7	22·6

Angela decides to look at the sales figures of the major competitor.

Year	2006	2007	2008	2009	2010	2011	2012
Sales in £million	28·1	28·1	28·2	27·8	25·7	23·8	22·8

She feels a bit better now, but she knows she still has a lot of work to do.

How should she present the sales information to the board of directors?

What you need to know

1 Calculate the value of each small division on these scales.

a  b  c 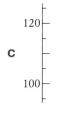 d 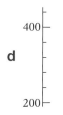 e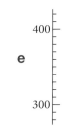

2 What is strange about these scales?

a 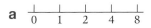 b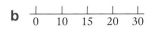

3 A Rubik's Cube is a puzzle. It is solved one layer at a time.

Several people try to solve the cube in an experiment.

A sheet records how well they perform.

The results are then transferred to a table.

The sheet is on the right:

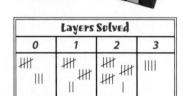

Layers Solved			
0	*1*	*2*	*3*
ⵜⵜ |||	ⵜⵜ ⵜⵜ ||	ⵜⵜ ⵜⵜ ⵜⵜ	||||

a Copy and complete the table of results.

Layers solved	0	1	2	3
Frequency	8			

b How many people took part in the experiment?

c If this group is typical, what is the most likely outcome if someone picked at random is given a Rubik's Cube to solve?

4 A small online mail order company records the cost of postage.

Cost	60p	90p	£1·20	£1·60
Frequency	5	10	15	2

State the mode and range of the postage cost paid.

6.1 Frequency tables

Frequency tables are a way of arranging data tidily so that we can see:

• the different categories examined

• how many pieces of data fall into each category.

Often a frequency table can be constructed from a record sheet where a list of the categories is made and a count is kept by means of tally marks.

Tally marks can also be used to help you sort any jumble of data into a frequency table.

Note that the tally marks are not part of the frequency table ... their use is just a device to get you from 'jumbled' to 'sorted'.

Example 1

Mark carries out a survey on the types of car that pass the school.

Renault	Renault	Vauxhall	Audi	Ford	Renault
Renault	Vauxhall	Vauxhall	Ford	Renault	Audi
Ford	Vauxhall	Renault	Ford	Audi	Ford
Ford	BMW	Ford	Audi	Vauxhall	Audi
Renault	Vauxhall	Vauxhall	BMW	Ford	Ford

a Use tally marks to help you sort the data into a frequency table.

b Draw the frequency table as it would appear in a report.

a Make a table with a separate column for type, tally and frequency.

Enter the different car types, counting and recording **one column of data at a time**.

Don't be tempted to scan the jumble of data for all the Renaults, then all the Fords, etc.

Here is the table with tallies recorded from the first two columns of data.

Type	Tally	Frequency
Renault	||||	
Ford	||	
BMW	|	
Audi		
Vauxhall	|||	

And here is the completed table.

Type	Tally	Frequency
Renault	ⅢⅡ ||	7
Ford	ⅢⅡ ||||	9
BMW	||	2
Audi	ⅢⅡ	5
Vauxhall	ⅢⅡ ||	7

b In a report you would omit the tally marks.

Type	Frequency
Renault	7
Ford	9
BMW	2
Audi	5
Vauxhall	7

Exercise 6.1A

1 Here is a frequency table showing the main courses chosen by people in a restaurant.

Main course	Frequency
Beef bourguignon	7
Baked sea bass	14
Scallops in cream sauce	13
Goats' cheese tart (v)	20
Aberdeen Angus steak	7

a How many people were surveyed?

b What is the most popular main course?

c What is the least popular main course?

d It is generally recognised that we should reduce the amount of red meat we eat.

Are the people in this survey heeding the advice?

2 The table records the reasons for absence from work at the dockside one month.

Reason for absence	Frequency
Common cold	9
Influenza (flu)	6
Upset stomach	10
Back injury	27
Other	8

a How many people were off work that month?

b How many more people were off with a back injury than any other single cause?

c What percentage of the absentees were off with flu?

d Which of the following steps would be the most effective in addressing the problem of absenteeism at the dock?

   A Advice on washing hands regularly.

   B Advice on the preparation of food.

   C Advice on lifting heavy objects.

3 Here is a list of the Champions League winners over the last 20 years.

1993 Olympique Marseille [F]   1994 AC Milan [I]   1995 Ajax Amsterdam [N]
1996 Juventus (Turin) [I]   1997 Borussia Dortmund [G]   1998 Real Madrid [S]
1999 Manchester United [E]   2000 Real Madrid [S]   2001 Bayern Munich [G]
2002 Real Madrid [S]   2003 AC Milan [I]   2004 Porto (Oporto) [P]
2005 Liverpool [E]   2006 Barcelona [S]   2007 AC Milan [I]
2008 Manchester United [E]   2009 Barcelona [S]   2010 Inter Milan [I]
2011 Barcelona [S]   2012 Chelsea (London) [E]

F = France, I = Italy, N = Netherlands, G = Germany, S = Spain, E = England, P = Portugal

a Make a frequency table counting the Champions League wins for **each country**.

b Which country had the most winners in this time period?

c What fraction of the years had an English winner?

4 For a national survey on ornithology, Mark is recording the birds that visit his bird table.

blackbird   robin   dunnock   blackbird   chaffinch
chaffinch   chaffinch   blackbird   chaffinch   chaffinch
wood pigeon   blue tit   chaffinch   coal tit
chaffinch   chaffinch   blue tit   blackbird   siskin
dunnock   chaffinch   chaffinch   greenfinch   robin
coal tit   wood pigeon   dunnock   robin

a Make a frequency table of the species of birds visiting the table.

b What was the most frequent visitor?

c What fraction of the visits were made by blackbirds?

Example 2

Grace and Torrance were planning a holiday in Skye.

They looked on the internet for a holiday cottage with one bedroom.

Here are the prices per week in pounds.

295	335	350	350	340	295
315	360	195	330	280	390
380	349	340	295	335	280
360	395	350	399		

a Why would it be impractical to have a frequency table that contains every price from £195 to £399?

b Make a frequency table by deciding in which of the following intervals each price lies. [£p is the price per week]

$$150 \leqslant p < 200 \quad 200 \leqslant p < 250 \quad 250 \leqslant p < 300 \quad 300 \leqslant p < 350 \quad 350 \leqslant p < 400$$

c What is the most common class interval for a price to lie?

a There would be over 200 price categories and most of them would have a frequency of zero. Using class intervals allows us to spot patterns and trends easier.

b We need to be careful as we enter the data:
£350 belongs to $350 \leqslant p < 400$ and not to $300 \leqslant p < 350$.

Price (p) in £	Tally	Frequency
$150 \leqslant p < 200$			1
$200 \leqslant p < 250$		0
$250 \leqslant p < 300$	⅏	5
$300 \leqslant p < 350$	⅏ ||	7
$350 \leqslant p < 400$	⅏ ||||	9

c The most common price range is $350 \leqslant p < 400$.

Exercise 6.1B

1 Here is the number of goals scored by Scotland in their games between September 2009 and February 2013.

2 0 0 0 1 0 0 2 0
2 3 3 0 3 0 2 1 1
1 2 1 1 3 0 1 0 1

a Make a frequency table for this information.

b What is the modal number of goals scored in a game?
(Remember: the mode is the most frequent.)

2 Sandy's boss often gives him the chance to work overtime.

He keeps a record of how many hours overtime he works in a week.

Here is a list of the number of hours of overtime that Sandy worked each week from March till his summer holidays in August.

     5 5 5 6 7 10 10 9 10 8 10

      10 7 9 8 10 7 7 9 8 10 10

a Make a frequency table to record Sandy's overtime.

b For what fraction of the weeks did he work more than 7 hours' overtime?

3 Julia is a gymnast. She regularly competes in competitions.

Here is a list of her position in the last 25 competitions she entered.

     5 4 7 3 5   5 6 4 5 6   5 3 5 7 8   8 7 6 4 4   4 3 3 4 4

a Put this information into a frequency table.

b What is her modal position?

c What was her best position?

d What was her worst position?

e What was the median of her positions?
   (Remember: the median is the middle score when the list is put in order, lowest to highest.)

4 Here are the percentage scores for the students in Mrs Higgins's class who sat a test on Pythagoras' theorem.

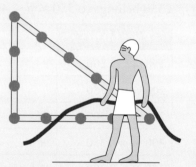

    82 73 77   75 79 83   85 95 71

    45 85 92   94 87 64   97 93 94

    86 83 81   63 85 92   78 72 61

a Sort the data into a frequency table using class intervals of

$40 \leqslant x < 50, 50 \leqslant x < 60$, etc., where $x\%$ represents the score.

b Which interval is the modal class?

c Mrs Higgins said those who scored more than 79% 'knew their Pythagoras'.
   What fraction of the class was she talking about?

d Rounding the percentages to the nearest whole number, she decided that the bottom 10% would be awarded an E, the next 20% a D, the next 40% a C, the next 20% a B and the top 10% an A.
   i How many students achieved each grade?
   ii Can the students be identified from the frequency table?

6.2 Bar charts and line graphs

We can use a bar chart or a line graph to display the information in a frequency table.

A bar chart doesn't tell the reader any more than a frequency table, but it does make it much easier to understand the information 'at a glance'.

When we take the data from a table and construct a graph we should be careful not to lose information.

A person should be able to understand the context from the graph just as from the table.

So we need scales and units, labelling of axes and a title.

Line graphs are very helpful for seeing how one variable is changing as another variable is changing.

When there is a general pattern to this change we call it a **trend**.

We have to be careful with line graphs.

Sometimes it makes sense to estimate values between the plotted points (continuous data), but at other times it doesn't (discrete data).

Example 1

For a TV quiz show, 100 members of the public were asked to name a 'First Minister' of the Scottish Parliament.

The results were gathered in a frequency table.

First Minister	Frequency
Donald Dewar	31
Henry McLeish	3
Jack McConnell	16
Alex Salmond	50

Construct a bar chart to illustrate this information.

Note that the bar chart carries all the essential information.

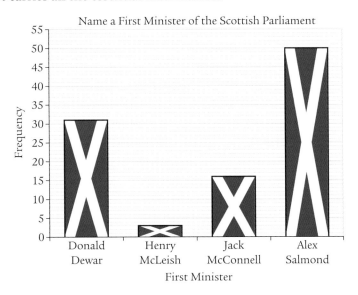

Example 2

The following table shows the temperature of a fridge measured every 30 minutes.

Time	08 00	08 30	09 00	09 30	10 00	10 30	11 00	11 30	12 00
Temp. (°C)	2·1	2·4	2·8	3·1	3·0	2·7	2·3	2·0	1·7

a Illustrate how the temperature of the fridge changes with time using a line graph.

b The temperature wasn't recorded at 10 45. Estimate the temperature at this time.

a

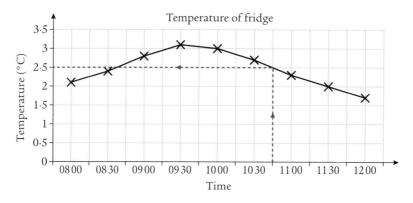

Temperature of fridge

b In this case the lines between the points give us a reasonable estimate of the temperature. (Temperature and time are continuous variables.)

So reading the graph, the estimated temperature at 10 45 is 2·5 °C.

Exercise 6.2A

1 Lesley was carrying out some market research for a chocolate manufacturer.

With her clipboard on a busy high street, she asked pedestrians,
'What is your favourite chocolate biscuit?'

Make a bar chart to display the data she collected.

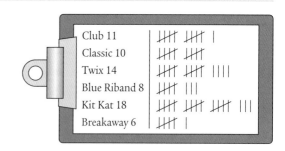

2 A holiday company carried out a survey on favourite destinations.

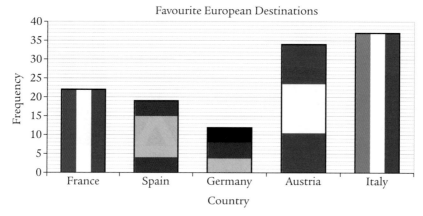

Favourite European Destinations

a Arrange the countries in order, highest frequency first.

b Which was the most popular country?

c Which was the least popular country?

d Which country got as many 'votes' as France and Germany put together?

e Which countries got more 'votes' than Spain and Germany put together?

3 Alistair drew a line graph to show how his football team's position in the league changed as the season progressed.

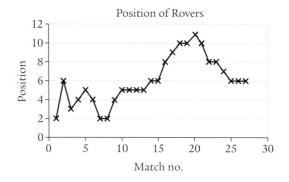

a What league position did Rovers have after:

  i 1 match

  ii 15 matches

  iii 25 matches

  iv 10 matches

  v 24 matches?

b Describe the club's progress between match 8 and match 20.

c Describe the club's progress between match 20 and match 25.

d Comment on the club's position between match 6 and 7.

4 Sharon is a self-employed hairdresser.

She keeps a close check on how well her business is doing.

The graph shows her profits at each year-end from 2006 to 2012.

a Describe the trend in Sharon's profits.

b In which year did Sharon's profits rise the most?

c In 2008 and 2010 Sharon ran advertising campaigns to try to attract more customers.
Do you think these campaigns were successful?

5 Craig played golf every Saturday.

He charted his progress weekly.

Month	Feb	March					April				May				
Date	22	1	8	15	22	29	5	12	19	26	3	10	17	24	31
Score	80	81	79	78	75	76	75	73	72	70	70	71	68	70	70

a Draw a line graph to show how Craig's score changes as the season progresses.

b Describe the general trend in Craig's scores.

c Name four Saturday's that went against the general trend.

d Usually golfers improve as the season goes on.
Is this the case with Craig? Remember, the lower the score the better!

6 A local museum keeps attendance figures for each month throughout the year.

Jan	Feb	Mar	Apr	May	Jun	Jul	Aug	Sep	Oct	Nov	Dec
2300	1800	1900	4300	3700	4100	6200	7300	4600	4800	2500	2100

a Draw a line graph to illustrate this information.

b Describe the trend in the attendances from February to August.

c Which month goes against the trend?

d Describe the trend from August to December.

e Which month goes against the trend?

f The following year the museum had an attendance of 2500 in February.
Can you make a comment on whether this is good, bad or indifferent?

Example 3

John compared the shoe sizes of boys in S1 with boys in S3.

S1 Boys	
Shoe size	**Frequency**
6	2
7	4
8	13
9	6
10	3

S3 Boys	
Shoe size	**Frequency**
8	1
9	5
10	12
11	7
12	3

Draw a **comparative** bar chart of the S1 and S3 shoe sizes.

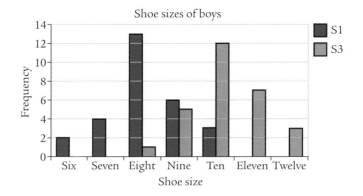

Note that since both sets of figures are plotted on the same chart, a key or legend is required.

Exercise 6.2B

1 In *TV Talent*, a talent show, the votes for the five acts were as follows:

Big Muddle 350 000

Lily Girl 250 000

Freddie Coz 150 000

Art James 225 000

Holly Blurs 175 000.

Using axes like these shown, construct a bar chart to display this information.

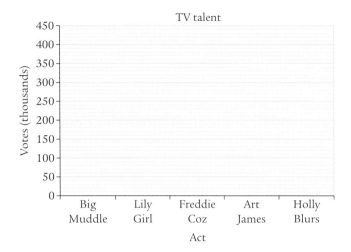

2 A holiday brochure shows a bar chart that compares the average maximum daytime temperature in Majorca to the average maximum daytime temperature in Edinburgh.

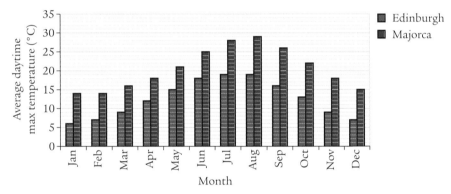

a Which month is the warmest in Majorca?

b Which two months are the warmest in Edinburgh?

c In which month is there the biggest difference in temperature between Edinburgh and Majorca?

d What is the smallest difference in temperature?

e In which month(s) does this occur?

f How can you tell from the chart that Edinburgh and Majorca are both in the same hemisphere on Earth?

3 The average mark of boys in their maths exam is compared to that of girls.

In 2007, the headteacher pointed out to the head of maths that the girls' average mark was trailing 5% behind the boys' average mark.

He asked the head of maths to try to make sure that the girls' performance improved.

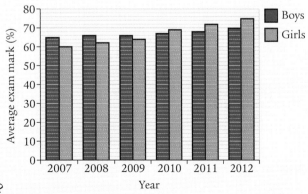

a Was this achieved?

b In which year did the girls overtake the boys?

c In 2012, the headteacher said to the head of maths, 'I'm very concerned that the boys are now trailing 5% behind the girls. I can't believe that the boys' performance can have declined so much in six years.'

If you were the head of maths, what would you say to the headteacher to reassure him about the boys' performance?

d If you wished to illustrate the difference between boys and girls, it may be better to number the y-axis from 60% to 75%.
  i Draw the comparative bar chart using this new scale.
  ii Does it highlight the difference as hoped?

4 The average daily hours of sunshine in Glasgow and in Sydney, Australia are compared.

Month	Jan	Feb	Mar	Apr	May	Jun	Jul	Aug	Sep	Oct	Nov	Dec
Sydney	9	8	8	8	7	6	7	8	8	9	9	9
Glasgow	1	2	3	5	6	6	5	5	4	3	2	1

a Draw a line graph to illustrate the comparison.
b What feature of the graph tells you that the two locations are in different hemispheres of the Earth?
c What feature tells you that Sydney is nearer to the equator than Glasgow?

6.3 Pie charts

A sense of proportion

When we wish to consider the fraction that falls into a particular category, then the pie chart is the most appropriate form of diagram.

The area of each section of pie is proportional to the frequency and thus, given that the pie is circular, is proportional to the angle at the centre. However, beware of representing the data in a 3-D diagram, as here area is not proportional to the angle at the centre. The 3-D effect diagram does not fit this description.

For example, at a car plant a paint problem appeared on some models. The problem seemed to be related to the colour of the car. A quick survey produced the following table.

Test colour	Frequency of flaw
Light blue	6
Red	8
Yellow	4
Green	10
Dark blue	4

An analyst tried to illustrate the data in a pie chart. Note that a 3-D diagram does not give an accurate picture. The angle at the centre of each segment is either too big or too small.

Look at the two pie charts on the following page and notice that the red area, which should occupy one-quarter of the pie, looks as if it occupies one-fifth of the 3-D diagram.

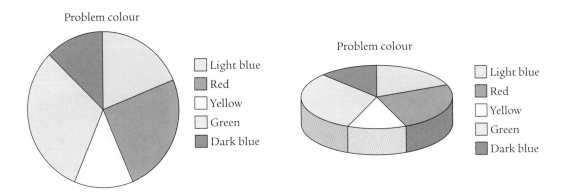

Problem colour

Problem colour

- ☐ Light blue
- ☐ Red
- ☐ Yellow
- ☐ Green
- ☐ Dark blue

Example 1

The students in Class 3M3 were asked how they came to school.

Their replies showed that 9 pupils walked to school, 8 came by bus, 4 cycled and 3 came by car.

Draw a pie chart to illustrate this data.

First we complete a frequency table with two added columns:

'Relative frequency' ... we calculate the fraction of the sample that falls within each category

'Angle at centre' ... we calculate that fraction of 360°.

We can now use the calculated angles to draw our pie chart.

Note that a key, or legend, is essential.

Mode of transport	Frequency	Relative frequency	Angle at centre
Walk	9	$\frac{9}{24} = \frac{3}{8}$	$\frac{3}{8}$ of 360° = 135°
Bus	8	$\frac{8}{24} = \frac{1}{3}$	$\frac{1}{3}$ of 360° = 120°
Cycle	4	$\frac{4}{24} = \frac{1}{6}$	$\frac{1}{6}$ of 360° = 60°
Car	3	$\frac{3}{24} = \frac{1}{8}$	$\frac{1}{8}$ of 360° = 45°

How do 3M3 get to school?

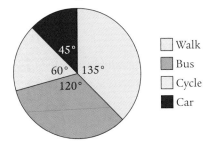

- ☐ Walk
- ☐ Bus
- ☐ Cycle
- ■ Car

Exercise 6.3A

1 Of the 15 players in Dukeston High School rugby team, 3 are in S4, 7 are in S5 and 5 are in S6.

   a Organise the data in a frequency table.

   b Add columns for: i 'Relative frequency'   ii 'Angle at centre'.

   c Show the composition of Dukeston High rugby team in a pie chart.

2 A food company making shortbread has to display the contents on the side of the packet. They use a pie chart to highlight the proportions.

   The shortbread is made of 600 grams of flour, 200 grams of sugar and 400 grams of butter.

   a Organise the data in a frequency table.

   b Show the composition of the shortbread as a pie chart.

3 A farmer hires a scanning machine to check if his ewes are pregnant.

Using the machine he can also tell how many lambs each ewe is carrying.

The more lambs that a ewe is carrying, then the more feeding the ewe will require.

Of 400 ewes scanned, the farmer finds that he has 12 ewes with no lamb, 103 ewes with one lamb, 259 ewes with two lambs and 26 ewes with three lambs.

Create a pie chart to illustrate this information. (You may find it useful to complete a frequency table and to round off your angles to the nearest degree.)

4 Researchers for a teenage magazine carried out a survey.

120 teenagers were asked what worried them most.

48 said their appearance, 33 said they worried about relations with friends, 24 said studies and exams, 9 said friends, 6 said family.

The editor wants to illustrate the findings on a pie chart.

Make a pie chart for the editor.

Example 2

In a survey, 1440 people were asked, 'Apart from a card, how do you send your seasonal greetings?'

Their answers are displayed using this pie chart. (The angles at the centre are given in degrees.)

a At a glance what was:

  i the most popular method

  ii the least popular method?

b How many people used:

  i the most popular method

  ii the least popular method?

c What was the range in frequencies?

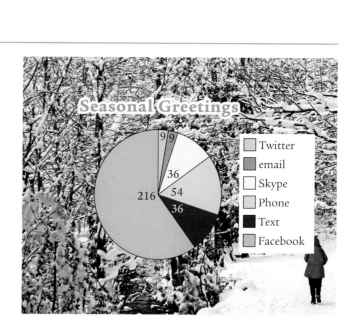

a **i** Facebook

  ii Twitter and email.

b **i** The fraction who said Facebook was $\frac{216}{360}$ (216° out of 360°).

   So the number who said Facebook $= \frac{216}{360}$ of $1440 = \frac{216}{360} \times 1440 = 864$ people.

  ii Twitter or email: $\frac{9}{360}$ of $1440 = \frac{9}{360} \times 1440 = 36$ people.

c Range $= 864 - 36 = 828$ people.

Exercise 6.3B

1 This pie chart shows the results of a survey to find out which platforms people prefer when gaming. (Angles are given in the diagram in degrees.)

A total of 270 people took part in the survey.

a What fraction of the sample chose the Wii?

b How many chose the PS3?

c How many more chose the Xbox than the PS2?

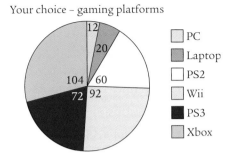

Your choice – gaming platforms

PC
Laptop
PS2
Wii
PS3
Xbox

2 Greg was trying to control his food intake.

He made a pie chart to illustrate the proportions of the different food groups (carbs, fat and protein) he had consumed at his dinner.

The total weight of the meal was 330 grams.

a For each sector, calculate the angle at the centre.

b Calculate the weight consumed of:
  i carbs   **ii** fat   **iii** protein.

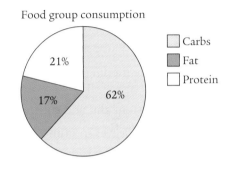

Food group consumption

Carbs
Fat
Protein

3 This pie chart shows how the electricity used in Britain in 2011 was generated.

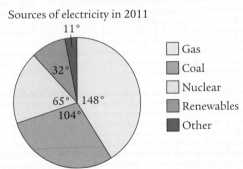

Sources of electricity in 2011

Gas
Coal
Nuclear
Renewables
Other

a What percentage of the electricity was generated by: **i** nuclear fuel **ii** coal?

b All the electricity generated goes into the National Grid. So in theory, the electricity you use comes from all these sources.

If your bill is £540, how many pounds worth of electricity did you use that was generated by:
  i coal   **ii** renewables?

6.4 Stem-and-leaf diagrams

A stem-and-leaf diagram is an efficient way of sorting data into numerical order.

This makes it very useful for finding the median and mode.

Example 1

Going to the football?

'How much does it cost for match tickets for one adult and two children, a match programme, three pies and drinks at Scotland's Premier League football clubs?'

Here are the results of a survey.

£48 £64 £53 £42 £68 £58

£56 £58 £43 £39 £47 £36

a   Put this information into a sorted stem-and-leaf diagram.

b   Find the median price.

c   What is the modal price?

a   Make the stem and add the leaves as you find them ... semi-sorted.

Arrange the leaves in numerical order ... sorted.

Cost of Premier League Football
3 | 9 6
4 | 8 3 2 7
5 | 6 8 3 8
6 | 4 8

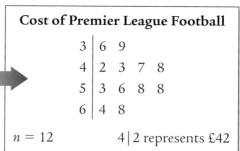

Cost of Premier League Football
3 | 6 9
4 | 2 3 7 8
5 | 3 6 8 8
6 | 4 8

$n = 12$ 4 \| 2 represents £42

b   There are 12 entries in the diagram so the median price is the mean of 6th and 7th entries.

Median price $= \dfrac{48 + 53}{2} = \dfrac{101}{2} = $ £50·50.

c   Since there are two occurrences of £58 and only one of every other price, the modal price = £58.

Exercise 6.4A

1   Thirty people attend a wedding. Their ages are noted.

25 27 24 24 23 47

51 53 50 49 22 23

24 23 23   3   7 14

15 14 81 73 72 65

27 32 41 25 23 22

a   Sort the information using a stem-and-leaf diagram.

b   What is the median age?

c   What is the modal age?

2 Jock was looking online for a new microwave oven.

Here are the prices (in £) of the microwaves he found.

40 55 68 55 73 63

71 66 89 69 53 79

71 64 66 78 73 52

53 78 85 91 70

a Create a stem-and-leaf diagram to sort this information.

b What is the median price?

c What is the range in prices?

d What percentage of the microwaves cost more than £75?

3 The players in Silver City's first team squad had their heights measured in centimetres.

185 187 182 175 191 187 186 178

183 184 180 190 182 179 186 185

a Make a stem-and-leaf diagram from this data.

b State the median height and range of heights.

4 In a biology experiment, the heights of 20 plants are measured at the beginning of the week and then again at the end of the week.

The increases in height (in cm) are calculated and recorded.

2·3 2·8 1·5 1·7 0·9 3·2 2·2 2·3 2·3 2·7

3·1 3·2 1·7 2·5 2·3 2·2 2·7 1·8 2·3 2·4

a Sort this data using a stem-and-leaf diagram.

b Find the median and the modal increase in length.

Back-to-back

Stem-and-leaf diagrams can also be helpful for comparing data sets.

For this we can use a back-to-back stem-and-leaf diagram.

Example 2

Here are the weights in kilograms of the competitors in a men's and a women's mountain bike competition.

Men: 83 85 91 84 99 85 86 79 83 101 93 91 94 81 77 88 83 95 101

Women: 75 68 72 77 82 63 82 75 72 78 64 69 73 72 73 72 71 78 65

a Draw a back-to-back stem-and-leaf diagram.

b What is the median of each group?

c What is the range of each group?

d Make two statements which compare the men's and women's weights.

a

<table>
<tr><td colspan="3" align="center">Weight of competitor</td></tr>
<tr><td align="center">Women</td><td></td><td align="center">Men</td></tr>
<tr><td align="right">9 8 5 4 3</td><td>6</td><td></td></tr>
<tr><td align="right">8 8 7 5 5 3 3 2 2 2 2 1</td><td>7</td><td>7 9</td></tr>
<tr><td align="right">2 2</td><td>8</td><td>1 3 3 3 4 5 5 6 8</td></tr>
<tr><td align="right"></td><td>9</td><td>1 1 3 4 5 9</td></tr>
<tr><td align="right"></td><td>10</td><td>1 1</td></tr>
</table>

$n = 19$ 7 | 5 represents 75 kg $n = 19$

b There are 19 entries in each group so in each case the median will be in the 10th position.

    Men's median weight = 86 kg, women's median weight = 72 kg.

c Men's range = 101 − 77 = 24 kg, women's range = 82 − 63 = 19 kg.

d On average the men are heavier than the women.

    The men's weights are more variable.

Exercise 6.4B

1 A cinema multiplex surveys the age of customers leaving the theatre.

    Screen 1 showed a cartoon, *Courage*.

    7 9 10 35 5 8 29 10 11 13 35 5 7 10 12

    42 6 9 12 45 71 10 12 73 14 15 17 8 9 29

    Screen 2 showed a James Blond movie, *Cloud Down*.

    18 20 45 41 51 48 25 25 33 32 47 42 27 21

    33 26 18 19 18 18 42 39 33 36 42 45 53 48

    a Draw a back-to-back stem-and-leaf diagram to help you compare the ages of the people who attended the films.

    b On average, which film attracted the younger audience?

    c Which film attracted a bigger spread in the age of the audience?

2 Mrs Brown ran special classes after school to help her students achieve better results in their exam.

    Here are the exam results (in %) of the students in Mrs Brown's class.

    Special class attenders: 76 69 81 73 54 35 77 74 76 65 68 77 64

    Non-attenders: 57 63 71 33 41 95 53 89 92 34 62 39 47 54 71 85 47

    a Make a back-to-back stem-and-leaf diagram.

    b In which group is the student who scores highest?

    c Find the median of each group.

    d On average, which group does better?

3 A carpet manufacturer did a survey to find out more about the size of living rooms.

Here are the sizes in square metres of 20 living rooms in houses that were valued at:

A £200 000 or over   **B** less than £200 000

    16·4 17·7 16·9 14·1 16·8   15·2 17·1 14·9 10·3 11·6

    19·2 19·5 18·3 17·5 18·7   15·5 17·4 10·5 10·4 11·1

    17·5 18·1 18·3 19·7 16·4   11·9 13·2 12·9 14·1 13·8

    16·5 17·6 14·6 17·8 17·4   13·0 12·6 14·6 17·3 12·7

a Make a back-to-back stem-and-leaf diagram.

b By looking at the diagram can you tell which group, on average, has the bigger living rooms?

4 Carry out a survey of your own by asking students in your school a simple question. Use techniques learned in this chapter to collect and compare the information.

Now use a spreadsheet such as Excel to produce different types of statistical diagrams for your data, and decide which type of diagram best illustrates your findings.

Preparation for assessment

1 Kirsten asked some students in her school to name their favourite female singer.

Here are her results.

a Organise the data in a frequency table.

b Use the information in the frequency table to construct a bar chart.

2 The line graph shows the attendance figures for a country park over three years.

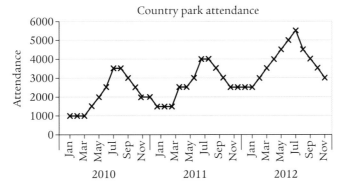

a What happened to the attendance between August and December in each of the three years?

b In which month in 2013 would you expect the attendance to be highest?

c In which month in 2013 would you expect the attendance to be lowest?

d In March of 2013, would you expect the attendance to be higher or lower than March 2012?

e The company managing the park claims that attendance is rising. Do you agree?

3   A tourist board, *Visitus*, was researching the places that
    people chose for a short-stay holiday.

    They asked 450 people to name their favourite
    short-stay destination in Scotland.
    Here are their findings.

    Aberdeen 50   Edinburgh 185   Glasgow 35

    Stirling 90   Inverness 70   Dundee 20

    By first finding the relative frequency of each city,
    draw a pie chart to illustrate the findings of the research.

4   Twenty-five people were asked the question, 'What age were you when you got married?'

    The results are shown in this stem-and-leaf diagram.

    > **Age when person got married**
    >
    > 1 | 8 8 9
    > 2 | 0 1 2 3 3 4 4 4 5 7
    > 3 | 0 1 1 3 4 5 5 7 9
    > 4 | 1 5
    > 5 | 1
    >
    > $n = 25$ 3 | 1 represents 31 years

    a   Find the median age.   b   What is the modal age?   c   What is the range of ages?

5   Remember Angela? She is in charge of sales at a large company.

    She is working in her office when the chief executive bursts in,

    'I'm really disappointed with our sales figures. Our sales have
    decreased for four years in a row.

    We were aiming to be ahead of our main competitor by 2014.

    I'll expect you to present these figures to the board of directors
    next Wednesday, and your presentation has to be good.'

    He hands her these sales figures for the last seven years.

Year	2006	2007	2008	2009	2010	2011	2012
Sales in £million	25·3	25·6	25·7	23·1	22·9	22·7	22·6

Angela decides to look at the sales figures of the major competitor.

Year	2006	2007	2008	2009	2010	2011	2012
Sales in £million	28·1	28·1	28·2	27·8	25·7	23·8	22·8

She feels a bit better now, but she knows she still has a lot of work to do.

How should she present the sales information to the board of directors?

Before we start...

In 1662, the physicist Robert Boyle confirmed that the pressure and the volume of a gas were related to each other.

By the early 1800s scientists found that the pressure, volume and temperature of a gas were related.

- a decrease in pressure causes a proportional increase in volume ... half the pressure, double the volume.
- a decrease in temperature causes a proportional decrease in volume ... half the temperature, half the volume.

On 14th October 2012, a base jumper named Felix Baumgartner ascended to a height of 39 km carried by a helium filled balloon. He then leapt from the balloon and began free-falling back to Earth.

As he fell he became the first human to break the sound barrier without any form of engine power.

The volume of helium in Baumgartner's balloon changed as he rose because of changes to the temperature and pressure.

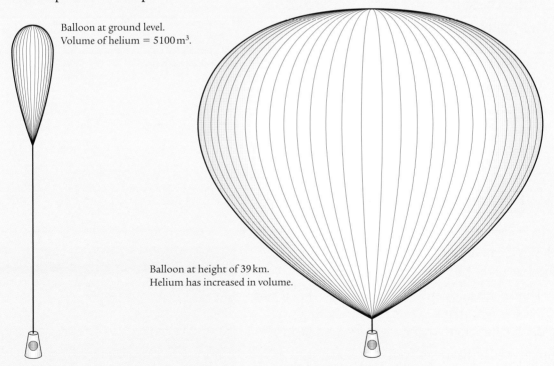

Balloon at ground level.
Volume of helium = 5100 m³.

Balloon at height of 39 km.
Helium has increased in volume.

Before take-off the helium had a volume of 5100 m³.

As the balloon rose the temperature decreased by a factor of 1·36 and the air pressure decreased by a factor of 226.

Can we find the volume of the balloon 39 km up in the stratosphere?

What you need to know

1   Express each ratio in its simplest form.

   a 5:10   **b** 9:6   **c** 12:20

2   Three boys are saving money for a holiday.
    In one month Adam's savings went from £200 to £250.
    Bill's went from £300 to £350.
    Charles' went from £400 to £500.
    Which two boys' savings increased in the same ratio?

3   One picture frame costs £12·99.
    How much will four cost?

4   A pack of eight cans of soft drink costs £3·20.
    How much is this per can?

5   The Jim Clark Rally takes approximately two hours to complete.
    Would this time increase, decrease or stay the same if:

   a   the distance covered in the rally was increased

   b   the average speed of the cars was slower?

7.1 Direct proportion

Two things are in **direct proportion** if they increase
and decrease in the same ratio.

Robert Hooke was a scientist who discovered that if you
doubled the weight that was hung from a spring you
would double the length that the spring stretched.

He would say that the weight and the stretch are
directly proportional.

> **Direct proportion:** an increase in one quantity will
> cause the related quantity to increase in the same
> ratio.

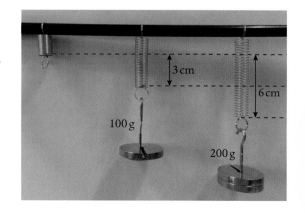

A greengrocer will tell you that if you want three times as many apples then you will pay three times as much. The cost and the quantity bought are in direct proportion.

When solving problems that involve direct proportion, say, finding the cost of 27 books, it makes sense to begin by finding the **cost of one book**.

Example 1

Karen normally works 30 hours per week and is paid £210.

Last week she only worked for 23 hours. How much will she earn?

We want to find out how much she'll be paid for 23 hours.

So, first find out how much she is paid for one hour.

In 30 hours she earns £210.

$\Rightarrow$ In one hour she earns $\dfrac{£210}{30} = £7$.

$\Rightarrow$ In 23 hours she will earn $£7 \times 23 = £161$.

Example 2

On a school trip three teachers are required to supervise up to 24 pupils.

How many pupils could go if there were 10 teachers?

We want to find out how many pupils can go if there are 10 teachers.

So, first find out how many pupils can go if there is one teacher.

Three teachers are needed to take 24 pupils.

$\Rightarrow$ One teacher is needed to take $\dfrac{24}{3} = 8$ pupils.

$\Rightarrow$ With 10 teachers we can take $8 \times 10 = 80$ pupils.

Exercise 7.1A

1 A phone call on John's mobile phone costs 5 pence per minute.

The cost and the duration of the call are in direct proportion.

a Copy and complete the table below.

Length of call, L (minutes)	1	2	3	4	5	10
Cost, c (p)	5					

b Explain how you calculate the cost of the call if you know the length of the call.

c **i** What happens to the cost if the call length doubles?

    ii Write down a formula for c when L is known.

2 Kirsten wants to work as a personal assistant when she leaves college.

In 8 minutes she's able to type 344 words.

a How many words can she type in one minute?

b If she can maintain this rate, how many words can she type per hour?

c Write down a formula that can be used to calculate W, the number of words typed when t hours is the length of time she's been typing.

3 Sachin knows that his car will use five litres of petrol to travel 45 miles.

  a How far can he travel using one litre of petrol?

  b Calculate how far he can travel on a tank of petrol holding 52 litres.

  c A formula connects petrol consumption, P litres, and distance travelled, d miles.

    It takes the form $d = kP$.

    What is the value of k?

4 To park for three hours in Edinburgh city centre it costs £10·80.

  a Calculate the cost of:

    i one hour's parking   **ii** 15 minutes parking.

  b A formula connects parking time, T hours, and cost, £c. It takes the form $c = kT$.

    i What is the value of k?

    ii What is the value of T when you park for 2 hours 20 minutes.

    iii Make T the subject of the formula.

5 On a rowing machine Greg can burn 160 calories in 10 minutes.

  a How many calories will he burn if he rows for 25 minutes at the same rate?

  b Write down a formula connecting c and t, where c is the number of calories burned when you row for t minutes.

  c How long would he need to row to burn 1000 calories?

6 Kassim changes £50 into US dollars and receives $80.

  a Assuming direct proportion, how many dollars could he get for £90?

  b When changing £120, Kassim was mistakenly given $200.

    i Was he given too much or too little?

    ii What was the difference between what he got and what he should have got?

7 The Steamship Society sells postcards and posters of the same picture.

The length and height of the postcard have been increased in proportion to produce the poster.

18 cm

12 cm

h cm

54 cm

The length of the picture increases from 18 cm to 54 cm.

  a By what factor has it been enlarged?

  b What is the height of the poster?

8 The pressure and temperature of a gas are directly proportional.

In an experiment, pressure was measured in kilopascal (kPa) and temperature in kelvin (K).

A sealed container of nitrogen at a temperature of 300 K has its pressure measured as 102 kPa.

a Calculate the pressure in the container when the temperature is raised to 400 K.

b At what temperature will the gas be when the pressure is 34 kPa?

c The formula $P = mT$ lets you find the pressure, P kPa when the temperature is T K.
What is the value of m?

Exercise 7.1B

1 Ally is building an ice fortress by making blocks out of compacted snow.

It takes him 18 minutes to make five blocks.

The finished fortress will need 75 blocks.

a How long will it take him to build the fortress?
Give your answer in hours and minutes.

b He's told that he can only stay out for one hour.
How many blocks can he make in that time?

2 Mary has friends in Australia that she likes to speak to on the phone.

When she calls at the weekend, a 20-minute conversation costs her £2·60.

a How much will a 45-minute conversation cost at the same rate?

b Calling a mobile phone is more expensive.

A 20-minute conversation would cost £5·80.

Calculate how much a 45-minute call to a mobile phone would be.

c What is the difference in price between calling the mobile phone for an hour and calling the home phone for an hour?

3 A nurse checks a patient's heart rate by counting their pulse for 10 seconds.

She then adjusts it to work out the number of beats per minute.

A normal resting heart rate is between 60 and 100 beats per minute.

a Calculate the heart rate in beats per minute of patients where, in 10 seconds, the nurse measures:

i 15 beats   **ii** 12 beats   **iii** 17 beats.

b A different nurse thinks you get a more accurate heart rate if you count the pulse for 15 seconds.

Calculate the heart rate per minute of these patients where, in 15 seconds, the second nurse measures:

i 20 beats   **ii** 16 beats   **iii** 19 beats.

4 A recipe for making eight chocolate cupcakes uses the ingredients shown.

Lynda needs to make 100 cupcakes for a church fête.

a Calculate how much of each ingredient she'll need.

b If, when making a batch of cupcakes, 420 g of margarine are used,

  i how many grams of self-raising flour are needed

  ii how many cupcakes are being made?

Ingredients　　　*Makes 8*
60 g soft margarine
50 g caster sugar
1 large egg
80 g self-raising flour
1 tablespoon cocoa powder
40 g plain chocolate

5 At the 2012 Olympic Games, Usain Bolt ran 100 m in 9·63 seconds.

a If he could maintain that pace how long would it take him to run the marathon (42 195 metres)?

Round your answer to the nearest whole second and change it into minutes and seconds.

b An assumption that time and distance are in proportion is being made here.

Comment on whether this is realistic.

c Consider the following in the same light.

  i It takes 10 minutes for one person to clean the inside of a car.

    How long will it take 10 people to do the same job?

  ii One person can run a mile in 4 minutes. How long would it take 4 men?

  iii The supermarket charges 10p for a carrier bag.

    One customer puts four oranges in his bag, another puts eight in his.

    Will the second customer pay twice as much as the first?

  iv Would you expect the cost of a taxi ride to be in direct proportion to the distance travelled?

6 Milk for babies can be made up by mixing a powdered formula with water.

To make milk for a newborn baby you add 15 g of formula to 90 ml of water.

As the baby grows they will need more milk.

a How much powder should be added to 150 ml of water?

b The water is boiled to kill all germs.

How much water needs to be boiled if you intend to use 35 g of formula?

7 Julie has an electricity bill for the period from 29th September to 5th January.

a How many days does this cover?

b The total cost of this bill is £116·01.

Assuming she uses roughly the same amount of electricity all year round, how much should her electricity come to in a year?

c Julie pays for her electricity by paying a direct debit of £35 per month.

  i Is this going to be too much or too little?

  ii By how much is it different?

Electricity Readings

Period	Previous	Present	Units used
29 Sep 12 to 5 Jan 13	05291	06097	806

Electricity Charges

806 units at 14·95p per unit	£120·49

Total Charges

Electricity	£120·49
Direct Debit Discount	£10·00
Sub-Total	**£110·49**
VAT at 5%	£5·52
Total Due	**£116·01**

8 Gold is one of the heaviest metals: 100 cubic centimetres of gold weigh 1930 grams.

   a How heavy would a 51·8 cm³ block of gold be?
     Round your answer to the nearest gram.

   b What is the volume of a one-gram block of gold?

   c The **density** of a material is usually given in grams per cm³, i.e. the weight of 1 cm³ of the material.

     Work out the densities of these substances in g/cm³.
     Give your answers to 2 decimal places.
       i Gold: 100 cm³ weighs 193 g.
      ii Lead: 40 cm³ weighs 452 g.
      iii Nickel: 50 cm³ weighs 445 g.
      iv Lithium: 70 cm³ weighs 37 g.
      v Water: 1 cm³ weighs 1 g.
      vi Cork: 10 cm³ weighs 2 g.

   d An object will float in a liquid if it is less dense than the liquid.

     Which of the substances in **c** float in water?

7.2 Graphs of direct proportion

If two quantities are in direct proportion, their graph is a straight line that passes through the origin.

In certain contexts such a graph can be used as a ready-reckoner.

For example, nurses can use one to convert a baby's weight from kilograms into pounds.

When you go abroad it can be useful to have one that converts between currencies.

Example 1

A joiner picks up an old DIY book from a second-hand bookshop.

All the measurements are given in inches.

He would like a quick way to convert.

He knows two things:

0 inches is equal to 0 cm; 50 inches is equal to 127 cm.

a Draw a graph that will change any length in the range 0 to 50 inches into centimetres.

b Use the graph to change: **i** 10 inches to cm **ii** 1 m to inches.

a

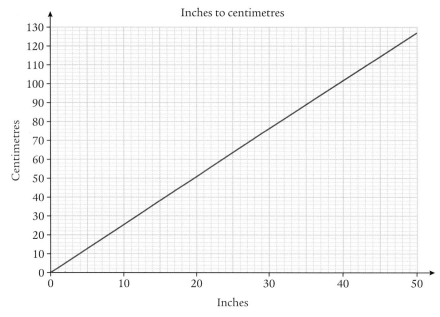

Inches to centimetres

b **i** From the graph, 10 inches looks to be about 25 cm.
ii From the graph, 1 m converts to almost 40 inches.

Exercise 7.2A

1 Use the graph shown in Example 1 to change:

a 30 inches to cm

b 5 inches to cm

c 60 cm to inches

d 10 cm to inches.

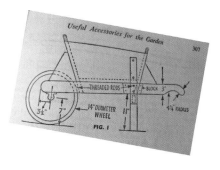

2 Imperial measurements, such as inches, feet, yards, miles and pints, are still quite common in the UK. Some other imperial units, such as furlongs, chains and leagues, are now used only in special situations, e.g. horse racing, if at all.

There are 3 feet in 1 yard (and 0 feet in 0 yards). From this we can also work out that there are 30 feet in 10 yards.

a Use this information to draw a graph that changes feet to yards.

b Use your graph to help change the following distances:
i 9 feet to yards   **ii** 24 feet to yards   **iii** 8 yards to feet.

3 In cricket, wickets are one **chain** apart. A chain is an old measurement equal to 22 yards.

Five chains are the same as 110 yards.

a Draw a graph to change chains to yards for distances up to five chains.

b Use your graph to help change the following distances:
i 3 chains to yards   **ii** 88 yards to chains   **iii** 55 yards to chains.

   c Horse racing still uses the **furlong** as the unit of length.

      One furlong = 10 chains.

       i Make a conversion graph for chains to furlongs.
      ii The course of the Grand National is 36 furlongs in length. Express this in yards.

4 When changing pounds to Australian dollars, £50 is roughly equivalent to $75.
   We also know that £0 = $0.

   a Draw a graph that will change up to £50 into Australian dollars.

   b Change the following amounts:
       i £20 to dollars   ii £35 to dollars   iii $63 to pounds.

   c Estimate the value, in Australian dollars, of:
       i £200   ii £350   iii £575.

Exercise 7.2B

1 Kilograms and pounds are two ways to measure mass.

   A mass of 15 kg is approximately 33 pounds.

   a Draw a graph that will change masses in the range 0 kg to 20 kg into pounds.

   b Change the following amounts:
       i 8 kg to pounds   ii 20 pounds to kg   iii 37 pounds to kg.

   c Estimate the mass, in pounds, of:
       i 200 kg   ii 1000 kg.

2 When changing pounds to US dollars, £50 converts to $80.

   a Draw a conversion graph to show this.

   b Convert: i £20 to dollars   ii $50 to pounds.

   c Estimate the value, in dollars, of:
       i £1250   ii £5750.

3 Peter decided to check his GPS in his car.

   He had planned a trip from Kirkcudbright to Aviemore.

   Using a route planner he found that the distance was 230 miles and the journey would take
   four hours.

   He assumed the distance travelled would be proportional to the time on the road, and drew a line
   graph to describe the journey.

   a Draw the graph that Peter drew.

   b How far from home does the graph say he will be when he has been on the road for:
       i 1 hour   ii 2 hours   iii 3 hours?

   c How long should he have been travelling when the car's odometer read:
       i 100 miles   ii 200 miles?

   d While he travelled, Peter noted his GPS said he had travelled 25 miles after 25 minutes.
      Was he ahead or behind schedule?

4 Old recipe books will use °F while new ones will use °C.

Some refer to the settings on the oven, e.g. gas mark 5.

    a Investigate the relation between these scales of the measurement of temperature.

        i Are °F and °C in direct proportion?

        ii What about 'gas mark' and °F and °C?

    b Scientists will often work in Kelvin.

        i Investigate the relation between the Kelvin scale and the other scales.

        ii Considering direct proportion, why is this scale better than either Celsius or Fahrenheit?

    c Other scales are also used. Explore the internet to discover them and their value.

7.3 Inverse proportion

If one quantity increases while another decreases by the same factor then we say the two quantities are in **inverse proportion**.

For example, a farmer has food for a herd of cows, enough to last the herd a fixed number of days.

If he were to double the number of cows in his herd, he'd halve the time the food stock would last.

Again, when solving such problems, it is sensible to consider the amount needed for one day.

Suppose there is enough food stock to last eight cows for five days, and we want to know how long the food stock will last 16 cows.

We should first find for how long the food stock would last one cow.

Example 1

A lottery prize is shared equally among those whose numbers match all of the numbers in the draw.

Five people came forward as winners and were set to win £600 000 each.

However, one of the tickets was discovered to be for the wrong day, so the prize was re-divided among the four actual winners.

How much will each winner receive?

We want to find out how the prize will be divided among four winners.

First we work out how much one person would win if there were no other winners.

Five winners get £600 000 each.

$\Rightarrow$ One winner would get £600 000 × 5 = £3 000 000.

$\Rightarrow$ If there are four winners then each one gets $\dfrac{\text{£3 000 000}}{4}$ = £750 000.

> **Inverse proportion:** an increase in one quantity will cause the related quantity to decrease in the same ratio.

Example 2

A relay team is raising money for charity by running the length of Hadrian's Wall. They'll divide the distance equally among them and will each collect sponsorship for their leg of the journey.

Seven people are in the team and should each complete 12 miles.

However, they decide this might be too far, so they ask three more people to join the team.

How far will each runner have to go now?

We want to work out how the distance will be divided among 10 people.

First, work out how far one person, on their own, would have to run.

Seven runners run 12 miles each.

$\Rightarrow$ One runner would have to go $12 \times 7 = 84$ miles.

$\Rightarrow$ 10 runners would each run $\dfrac{84}{10} = 8.4$ miles.

Exercise 7.3A

1 A teacher buys a box of sweets to give to her class on the last day of term.

   She works out that if all 20 pupils are there, they will each get six sweets each.

   However, on the last day of term only 12 pupils are present.

   How many sweets can each pupil receive?

2 At an average speed of 6 mph, Scott can run to school in 25 minutes.

   On a bike he can cover the same journey at an average speed of 15 mph.

   How long will this take?

3 Cameron and Claire share a job delivering flyers for local takeaways.

   It takes them four-and-a-half hours each to deliver to every house in their town.

   a Their friend Kyle wants to help with the job too.
      i How long would it take if they shared the job among the three of them?
      ii What are you assuming when working out your answer?

   b For delivering the flyers Cameron and Claire were due to receive £18 each.

      If they have to split the money equally with Kyle how much will they each get?

4 A school needs to replace the calculators in the maths department.

   There are two types they are interested in:

         A a cheap but functional scientific calculator for £4

         B a really good one for £6.

  a At £4 per calculator the department has enough money to buy 80.

  i How many of the £6 calculator could they afford to buy instead?

  ii How much money would they have left?

  b If they decide to buy 30 of the cheaper calculators, how many of the dearer ones can they buy?

5 A construction firm is building a new section of road.

The job must be completed in eight weeks, so they assign a team of 12 men to complete it.

As work is about to start there is a spell of cold weather, and the ice and snow make it impossible to get started until two weeks later than planned.

How many men should be assigned to the job now if the firm still wants to hit the deadline?

6 Shavaz is writing an essay for English.

If he types it in a font with size 15 he can fit 40 lines of writing on a page.

His teacher says that it will look more professional if he changes the font to size 12.

  a How many lines will fit now?

  b What assumption are you making in answering the question?

Exercise 7.3B

1 Microwave ovens come with different power ratings.

Food needs to be cooked for longer in microwaves with lower power ratings.

In a 800 W microwave oven a ready meal needs to be cooked for five minutes.

  a For how long should you cook the meal in a 1000 W microwave?

  b What assumption are you making in answering the question?

2 At a fixed temperature, the pressure of a given mass of gas is inversely proportional to its volume.

Pressure can be measured in kilopascals (kPa) and volume measured in cubic metres (m³).

For a particular mass of oxygen, when the pressure is 200 kPa, it has a volume of 0·8 m³.

Calculate the volume it would occupy if the pressure were increased to 250 kPa.

3 A flash drive can hold 2600 high-quality digital photographs, each of which requires 1·5 MB of memory.

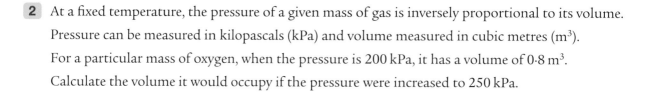

The photos can be compressed to 1·3 MB each.

  a If this is done, how many will fit on the flash drive?

  b 'Thumbnails' of the photos use 25 kB of memory each.

  There are 1024 kB in 1 MB.

  How many 'thumbnails' can be stored on the flash drive?

4 Angela has been given a gift card that enables her to download music and videos from an eStore.

There is enough credit on the card to allow her to buy 40 music tracks at 75p each.

The online store has selected music albums on special offer. These cost £5 each.

  a How many albums can she afford to get with her gift card?

  b **i** If she buys two albums, how many individual tracks can she get with her remaining credit?
    ii If the credit drops below 75p, then it is lost. How much is lost in this case?

5 In an electrical circuit, if you increase the resistance, then the current decreases proportionally.

A circuit is set up that has a current of 5 amps and a resistance of 26 ohms.

Calculate the current in the circuit when the resistance is changed to:

  a 23 ohms   **b** 30 ohms   **c** 18 ohms.

6 A company is investigating different sizes of drinks cans.

The cans must be cylindrical and hold a fixed volume of juice.

The cross-sectional area of a can is inversely proportional to its height.

The cross-section of the can they are using at the moment is 33·2 cm² and it has a height of 11·4 cm.

  a How tall would the can be if they changed the area of the cross-section to:
    i 33 cm²   **ii** 30 cm²   **iii** 35 cm²?
    Round your answers to 2 decimal places.

  b What would the cross-sectional area be if the height of the selected can was:
    i 15 cm   **ii** 19·45 cm?

7.4 Mixed problems

In these exercises you'll meet problems that may feature direct proportion, inverse proportion or other ways that two quantities may be related.

It's up to you to read each question carefully and decide whichever strategy will work.

Be careful, not everything is in proportion! Here are some classic situations:

It will take three minutes to boil an egg. How long will it take to boil three eggs?

Henry the Eighth had six wives. How many wives had Henry the Fourth?

If you put one stamp on an envelope it will weigh 4·1 grams. How much will it weigh if I put three stamps on it?

Example 1

A music teacher visits a school for a fixed time and divides her time equally among the 10 pupils who need tuition. Each pupil gets 24 minutes.

An extra two pupils join the school and they also would like some music tuition.

How much time will each of the 12 pupils get now?

We can treat this as a case of inverse proportion.

10 pupils get 24 minutes each.

$\Rightarrow$ 1 pupil would get $24 \times 10 = 240$ minutes.

$\Rightarrow$ 12 pupils would get $\dfrac{240}{12} = 20$ minutes each.

Note that we assumed inverse proportion was in play. However, at some point, the result would have been silly ... 120 pupils would get $\dfrac{240}{120} = 2$ minutes each.

You couldn't give a two-minute music lesson!

Example 2

At the weather station they measured 8 cm of rain over three days.

How much will they measure over five days?

Unless you are told that it rained constantly for a week, then you cannot assume that rainfall and time are in proportion.

Exercise 7.4A

1 Say whether each situation is likely to be an example of direct proportion, inverse proportion or not in proportion.

    a The number of people painting a house and the time taken to do it.

    b The time taken to mow a lawn and the total area of grass.

    c The cost of a piece of cheese and its weight.

    d The time taken to change a light bulb and the number of people helping.

    e The time taken to change every light bulb in the school and the number of people helping.

    f The age of a car and the number of wheels it has.

2 A company sells calculators to schools.

    For orders of fewer than 30 calculators they charge a set price per calculator.

    Additional calculators can be bought at a reduced price.

    a Make a copy of the grid shown.

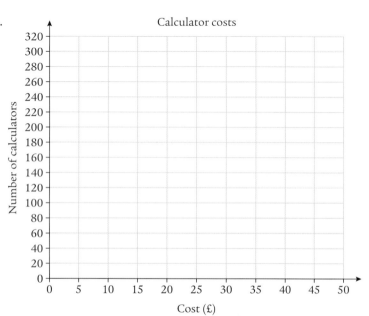

b No calculators will cost £0. Thirty calculators cost £210.

Draw a line connecting (0, 0) to (30, 210).

c If you buy 50 calculators you pay £310.
  i Plot a point on your grid to represent this information.
  ii Draw a straight line connecting this point to (30, 210).

If you keep in mind that you can only buy a whole number of calculators and that only some of the points on the line are sensible, you can use your graph to cost any amount of calculators up to a maximum of 50.

d Estimate the cost of orders for:
  i 20 calculators **ii** 35 calculators.

e How many calculators can be bought for:
  i £260 **ii** £100?

f Is the number of calculators purchased directly proportional to the cost?

Explain your answer in terms of the graph.

3 The international space station orbits the Earth 16 times in 24 hours.
  a How long would it take it to orbit the Earth five times?

  b An astronomer observed it at a particular point in the sky.
  At a time 200 hours later he looked for it again in the same spot.
  How long would he have to wait before it came into view?

4 A train can travel from Glasgow to Carlisle in 60 minutes, averaging a speed of 96 miles per hour.
  By car it takes 90 minutes to do the same journey.
  a Calculate the average speed of the car journey.
  b What assumptions did you make?

5 In the final of a 100-metre race there were eight competitors.
  How many competitors were there in the final of the 200-metre race?

6 A car rental firm hires out cars for a set fee per day.
  It costs £210 to hire a small car for 14 days.
  a How much will it cost to hire the car for six days?
  b **i** For how many complete days can the car be hired for £300?
      ii How much change would there be?

7   In March 2021 a census will take place to collect information about every person living in the UK. A team of people will be employed in each area to check that people have filled in forms correctly and to provide help if they need it.

For one town it has been estimated that it will take six days for a team of 25 people to visit every house. Note that when making estimates, proportion is often assumed.

   a   How many people would need to be employed if they wanted every house to be visited in five days?

   b   If a person is paid £70 a day to work on the census, calculate the wage bill:
   i   for the 25 people working six days
   ii   for getting the job done in five days.

   c   Comment on your answer to **b**.

Exercise 7.4B

1   Andy is a plumber and charges his customers a call-out fee plus an hourly rate.

If he is called to a job that takes two hours he will charge £50.

If he goes to a job and it takes seven hours he charges £125.

   a   Draw a graph to show how much he will charge for jobs that take from 0 to 8 hours.

   b   What is Andy's call-out fee (how much does he charge for 0 hours work)?

   c   What does he charge per hour?

   d   What stops Andy's prices being in proportion to the time it takes him to do the job?

2   Martin organises weddings. He buys his wine in bulk from a wholesaler.

At £8 per bottle he can afford to buy 100 bottles of wine.

If he can negotiate the price down to £7·50 per bottle how many bottles can he afford?

3   Vijay is playing cricket. In the last three overs he has made 42 runs.

If he continues at this rate, after how many overs will he have made at least 100 runs?

4   A bus journey from Earlston to Galashiels is nine miles and costs £3·60.

A bus journey from Earlston to Melrose is six miles and costs £2·70.

Are the prices in proportion? Explain your answer.

5   The label on a bag of chocolate eggs says that 100 g contains 490 calories.

The bag contains 240 g. How many calories are in the whole bag?

6   A supermarket sells large multi-packs of toilet rolls.

A pack of 12 toilet rolls costs £4·92, 18 rolls cost £8 and 24 rolls cost £10.

   a   Are the prices of the toilet rolls in proportion? Explain your answer.

   b   The largest pack has a banner stating 'BIGGER PACK, BETTER VALUE'.

   Explain why this is false advertising. Which pack is the best value?

7 A distillery is bottling whisky that has been aged in large barrels for 18 years.

From five barrels they can fill 2000 standard-sized bottles.

  a How many standard-sized bottles could be filled from 12 barrels?

  b The standard-sized bottle is 750 ml.

   If they change to a larger one-litre bottle, how many bottles can they fill from their 12 barrels?

8 In 1843 the following puzzle appeared in the newspaper.

The problem uses 'old money'. You can use modern money.

  a At three for a penny, what should 30 apples cost?

  b At two for a penny, what should 30 apples cost?

  c How much is made in total on the 60 apples?

  d C argues that three for a penny and two for a penny works out at five for two pence. What do 60 apples cost at this rate?

  e What is the cause of the 1d. (one penny) difference?

> **ARITHMETICAL PUZZLE.**
>
> | | s. | d. |
> |---|---|---|
> | A, sells 30 apples in the market at three for a penny, receiving of course...... .... | 0 | 10 |
> | B, sells 30 at two a penny, receiving .... | 1 | 3 |
> | | 2 | 1 |
>
> C, having 60 apples, says he will sell them at the same average price, namely, five for two-pence, but, in doing so, he finds he receives only... 2 0
>
> What is the cause of the 1d. difference ?
>
> S. M. P.

7.5 Formulae

A formula is a good way of expressing the relation between variables.

You'll find them in science, home economics, technology and many more places.

They are essential when working with spreadsheets.

Whenever you use a formula you should follow the steps below.

Copy the formula.

Substitute numbers into the formula.

Evaluate the answer.

Check that the answer is reasonable.

Example 1

The cost of a call on John's mobile phone contract can be worked out using the formula $C = 5t$, where C pence represents the cost of the call and t minutes is the length of the call.

Find the cost of a call that lasts 12 minutes.

$C = 5t$

$\Rightarrow C = 5 . 12$ Here we use the dot notation: $5 . 12$ means the same as 5×12

$\Rightarrow C = 60$ pence

Example 2

In electrical circuits the voltage, current and resistance are linked by the formula:

$$I = \frac{V}{R}$$

where I amps is the current, V volts is the voltage and R ohms is the resistance.

Calculate I, when $V = 230$ and $R = 46$.

$$I = \frac{V}{R}$$
$$\Rightarrow I = \frac{230}{46}$$
$$\Rightarrow I = 5$$

Exercise 7.5A

1 A car hire firm uses the formula $C = 18d$ when someone wants to rent a medium sized car.
   The total cost to hire the car is £C and d is the number of days for which the car is hired out.
   a Show how the formula would be used to find the cost of car hire for two weeks.
   b How might the formula be adapted to allow you to calculate d when given C?

2 In a local lottery a monthly prize of £1000 is shared among the number of winners.
   The amount each winner gets can be worked out using the formula, $P = \frac{1000}{n}$, where £P is a winner's share and n is the number of winners.
   a Show how the formula would be used to calculate a share when four people win.
   b How might the formula be adapted to allow you to calculate n when given P?

3 For an equilateral triangle the formula $P = 3s$ can be used to find the perimeter, P units, of a triangle with sides s units long.
   a Use the formula to find the perimeter of an equilateral triangle of side:
      i 8 cm   ii 3 km   iii 5·5 mm.
   b Adapt the formula to allow you to calculate s when given P.

4 The following formulae appear in science and mathematics. You may have seen some of them before.
   Evaluate each formula for the values given.
   a $E = Pt$ Find E when $P = 1200$ and $t = 30$.
   b $s = vt$ Find s when $v = 27$ and $t = 50$.
   c $r = \frac{D}{2}$ Find r when $D = 16$.
   d $P = \frac{F}{A}$ Find P when $F = 30$ and $A = 0·04$.
   e $a = \frac{F}{m}$ Find a when $F = 392$ and $m = 40$.
   f $E = mgh$ Find E when $m = 80, g = 10$ and $h = 8$.
   g $C = 2\pi r$ Find C taking $\pi = 3·14$ and $r = 25$.
   h $v = u + at$ Find v when $u = 5, a = 2$ and $t = 10$.

Exercise 7.5B

1 An electrician charges a call-out fee of £20 plus £12 per hour for his work.

He uses a spreadsheet to calculate the fee:

<>	A	B	C	D
1	Call-out fee	Rate per hour	Hours	Total Charge.
2	20	12	3	56
3	20	12	2	
4	20	12	5	
5				
6				

a The total charge in D2 is calculated using the formula: = A2+B2*C2

Check that the value in D2 does work out as 56 when A2 = 20, B2 = 12 and C2 = 3.

b What is the equivalent formula that should be typed in:
   i D3   **ii** D4?

c What values should appear in
   i D3   **ii** D4?

d Row 5 is used to calculate a job that lasts 2 hours 30 minutes.
   i What is typed into each cell of row 5?
   ii What value appears in D5?

2 A flow diagram has been drawn up to help you work out the cooking time, in minutes, for a turkey, given its weight.

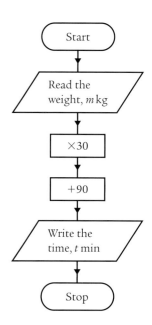

a Use the flow diagram to work out how long you would cook a 5 kg turkey for.
   Give your answer in hours.

b The formula $t = 30m + 90$ is another way to find the cooking time.
   Use the formula to calculate the cooking time for a turkey that weighs:
   i 2 kg   **ii** 7 kg.

3 Temperatures can be measured either in °Celsius (C) or in °Fahrenheit (F).

The formula for changing Celsius to Fahrenheit is $F = 1{\cdot}8C + 32$.

a Use the formula to change 20 °C into °Fahrenheit.

b Change 28 °C into °Fahrenheit.

c What temperature in °Celsius would give an answer of 32 °F?

4 The following formulae appear in science and mathematics. You may have seen some of them before.

Evaluate each formula for the values given.

a $A = \frac{1}{2}bh$ Find A when $b = 12$ and $h = 3$.

b $E = IVt$ Find E when $I = 10$, $V = 230$ and $t = 60$.

c $A = \pi r^2$ Find A when $\pi = 3.14$ and $r = 25$.

d $V = \frac{1}{3}\pi r^2 h$ Find V when $\pi = 3.14$, $r = 3$ and $h = 8$.

e $s = ut + \frac{1}{2}at^2$ Find s when $u = 8$, $t = 5$ and $a = 10$.

5 The diagram shows the net of a box that is to be made out of wood.

The area, A mm^2, of each rectangular face can be found by using the formula $A = lb$, given the length, l mm, and breadth, b mm, of the rectangle.

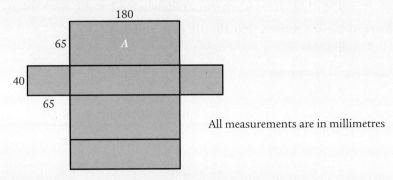

All measurements are in millimetres

a Find the total area of wood required to make the net.

b All six rectangles have been cut out with a tolerance of ± 2 mm.
   i What are the maximum values for the length and breadth of the rectangle A?
   ii What is the maximum area of A?
   iii What is the largest maximum possible area of the net?

Preparation for assessment

1 Say whether each scenario is an example of direct proportion, inverse proportion or no proportion.

   a The area of a patio and the number of paving slabs required.

   b The number of staff in a restaurant and the quality of the food.

   c The speed of a plane and the time taken to get to its destination.

   d The number of pages in a book and the number of words in it.

   e The speed someone can type and the time taken to retype an essay.

2 The force of gravity experienced by a satellite in orbit around the Earth is directly proportional to its mass.

   A satellite that has a mass of 500 kg experiences a gravitational force of 3500 N.

   Calculate the force experienced by a satellite with mass 800 kg.

3 The cost of a school trip is shared equally between all the students going on it.

If 40 students go, they will each have to pay £360.

How much must each student pay if 45 pupils go?

4 Marc is a dentist.

If he can do a check-up in 12 minutes, he can see 40 patients in his working day.

His colleague Tony takes 15 minutes per check-up.

How many patients can he expect to see in a working day?

5 a Use the formula $Q = CV$ to calculate the value of Q when $C = 3$ and $V = 8$.

b Use the formula $C = \dfrac{Q}{V}$ to calculate the value of C when $Q = 30$ and $V = 12$.

6 Remember Felix Baumgartner – the man who fell from space?

When dealing with gases their pressure, temperature and volume are related in such a way that:

- a decrease in pressure causes a proportional increase in volume
- a decrease in temperature causes a proportional decrease in volume.

The volume of helium in Baumgartner's balloon changed as he rose, because of changes to the temperature and pressure in the atmosphere.

Before take-off the helium had a volume of $5100\,\text{m}^3$.

As the balloon rose the temperature decreased by a factor of 1·36 and the air pressure decreased by a factor of 226.

Find the volume of the balloon 39 km up in the stratosphere.

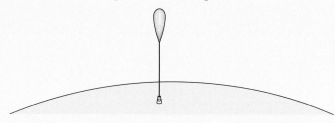

⏸ Before we start...

Margaret wanted to get some facts about the Ford Anglia 1960 model using old photos from the Museum of Transport.

All she knew was that the floor tiles were squares of side 310 mm (1 foot).

Is there enough information in the picture for her to estimate the size of the wheel base?

The wheel base is the distance between the two corresponding points on back and front wheels, for example, the centres or the points where the wheels touch the ground.

⊙ **What you need to know**

1 A company produces a puzzle cube.

The larger cube has an edge of length 6 cm.

  a What is the length of the edge of the smaller cube?

  b What is the area of:
    i a face of the larger cube
    ii a face of the smaller cube?

2 Find the length of the side of a square of area:

  a 49 cm²   **b** 1·21 cm².

3 A second puzzle cube has one face marked out like this:

The face is made from nine squares each of area 5 cm².

  a What is the area of the square ABCD ?

  b What is the length of AB?

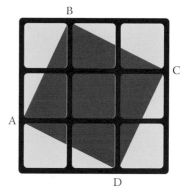

8.1 Practical geometry

The ancient Egyptian surveyors were known as 'rope stretchers'.

They used lengths of rope and stakes, the way we would use a set of compasses, to make shapes in the sand.

Example 1

Make a right angle using a straight edge and a set of compasses.

- Draw a line.

- Mark a point on the line and label it A.

- With your rope/compasses set to any distance, mark an arc cutting the line at a point B.

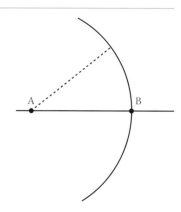

- Setting your rope/compasses to a shorter setting, draw an arc using B as your centre.

  The arcs should be big enough to cut each other above and below the line at points C and D.

- Draw a line between C and D.

  CD is at right angles to AB.

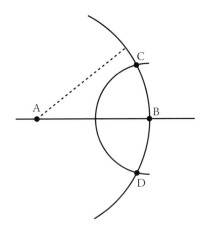

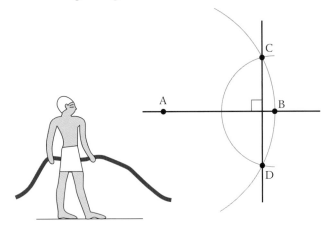

Exercise 8.1A

1 Consider the construction in Example 1.

    a What is the shape formed by joining ACBD?

    b How could you alter the instructions so that the line CD passes through the midpoint of AB?

    c How could you alter the instructions so that the line CD passes through a given point?

2 Using a straight edge and set of compasses only:

    a draw an angle of 60°

    b devise a set of steps for bisecting an angle

    c draw an angle of 30°

    d draw an angle of 45°.

3 **a** Using your compasses draw a triangle with sides 3 cm, 4 cm and 5 cm.
   You may use a ruler to set your compasses.

    b What is special about this triangle?

    c Why do engineers like to use triangles in their constructions?

Exercise 8.1B

1 All triangles that have sides 3 cm, 4 cm and 5 cm are congruent.

The Egyptian rope stretchers used a rope with 12 evenly spaced knots to do a very particular job.

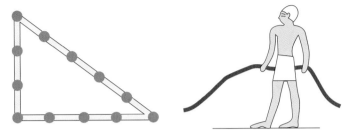

What might this job have been?

2 The ancient mathematicians explored number with the aid of pebbles.

A number was considered 'square' if that amount of pebbles formed a square, for example:

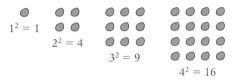

$1^2 = 1$ $2^2 = 4$ $3^2 = 9$ $4^2 = 16$

a List the first 15 square numbers.

b The number 324 is a square number.

How many pebbles are on each side of the square that can be made?

(This is called the square root of 324 and written $\sqrt{324}$.)

c Find the square root of the following:
  i 529   **ii** 2025   **iii** 10 000.

3 The Egyptians noticed something that tied together the points explored in Questions **1** and **2** above.

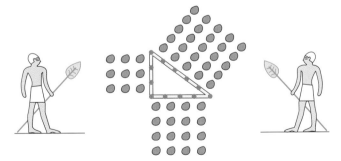

What do you notice?

8.2 Pythagoras' theorem

ABCD is a square. E, F, G and H are each positioned to be a units from one corner of ABCD and b units from the next corner clockwise.

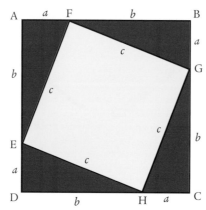

How do we know that the quadrilateral EFGH is a square?

By rotational symmetry each of the sides of EFGH are equal, say c units.

So the quadrilateral is a rhombus.

But by rotational symmetry, the four angles are equal ... and right angles.

So EFGH is a square.

So the yellow area is c^2 square units.

Now imagine that the four red triangles are rearranged to produce this second diagram.

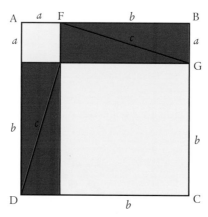

The yellow area in this second diagram is equal to $(a^2 + b^2)$ square units.

By the conservation of area, the red areas are equal in both diagrams.

So the yellow areas are equal in both diagrams.

So $a^2 + b^2 = c^2$

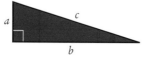

(Note: this worked only because the red triangle is right angled.)

This result is known as the Theorem of Pythagoras:

In any right-angled triangle, the square on the hypotenuse is equal to the sum of the squares on the other two sides.

If we know the lengths of two sides of a right-angled triangle, we can use the theorem to deduce the length of the third.

Example 1

Saltcoats is a town on the Ayrshire coast.
Its centre is 32 km west and 24 km south of the centre of Glasgow.

How far apart are the two town centres?

Sketch the Saltcoats/Glasgow triangle:

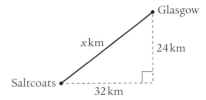

The triangle is right-angled ... so we can use Pythagoras' theorem:

$$x^2 = 24^2 + 32^2$$
$$\Rightarrow x^2 = 576 + 1024$$
$$\Rightarrow x^2 = 1600$$
$$\Rightarrow x = \sqrt{1600} = 40$$

Saltcoats is 40 km from Glasgow.

Example 2

A triangular buttress is built against a vertical wall to help support it.
It is 6 m tall and, at the bottom, comes out 2·5 m from the wall.

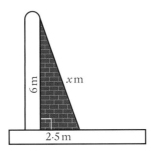

What is the length of the sloping edge of the buttress?

The triangle is right angled (vertical and horizontal)   ... so we can use Pythagoras' theorem:

$$x^2 - 2.5^2 + 6^2$$
$$\Rightarrow x^2 = 6.25 + 36$$
$$\Rightarrow x^2 = 42.25$$
$$\Rightarrow x = \sqrt{42.25} = 6.5$$

The sloping edge is 6·5 m long.

Exercise 8.2A

1 Calculate the hypotenuse in each case.

a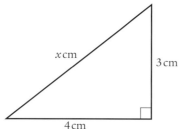

b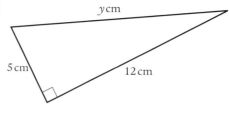

c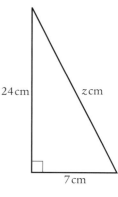

2 Calculate the *exact* length of each hypotenuse.

a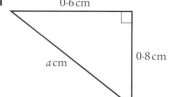

b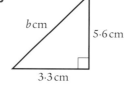

c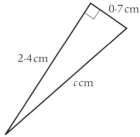

3 Calculate the length of the hypotenuse, correct to 1 decimal place, of a right-angled triangle whose shorter sides are:

  a 25 cm and 16 cm

  b 12·3 cm and 8·1 cm

  c 10 cm and 10 cm.

4 An electrician is wiring up a stall in an exhibition hall.

He must lead the wiring round a corner.
He has two choices:

Option 1: Start 10 m before the corner and continue 15 m after the corner.

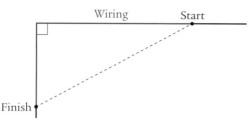

Option 2: Start 8 m before the corner with the same length of wiring and continue after the corner.

  a How far can he continue after the corner using the second option?

  b **i** In which option is the start further from the finish?

    ii By how much?

5 A rhone pipe takes waste water from the guttering in the roof to a drain.

It is shaped under the eaves of the house.

The pipe goes 320 mm in under the eaves for a drop of 250 mm.

Calculate the length of the pipe (AB) that is below the eaves.

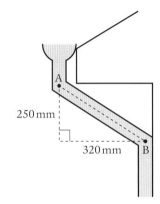

6 A telephone engineer puts in a new line to a house.

The new line is attached to the pole at a height of 8·5 m, and enters the house at a height of 4 metres.

It spans the 16 m gap without sag.

a Sketch a relevant right-angled triangle.

b Find the length of the new line.

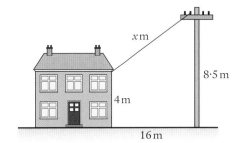

Exercise 8.2B

1 A TV mechanic brings an aerial cable into a rectangular room at an entry point.

He then has the choice of skirting the room bringing the cable to the TV via the red line, or going straight across the room, under the floorboards, via the broken line.

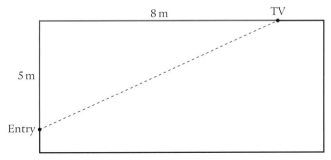

Cable costs £1 per metre (complete metres only). Labour is £10 per hour.

Being accessible, the red route will only take 1 hour's work.

Going under the floorboards is a one-and-a-half hour job.

a How much cable will be needed for the direct route?

b Which of the two routes is the cheaper? By how much?

2 A joiner is planning to make a lean-to shed.

He has to get an idea of its surface area to cost the project.

a Calculate the length of: **i** PQ **ii** PR.

b Use Pythagoras' theorem to find the length of QR.

c What is the area of the roof section?

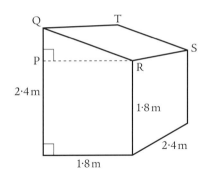

3 The same joiner is making garden furniture. He makes up a square of side 2 m, from which he means to cut the corners off to give him a regular octagon.

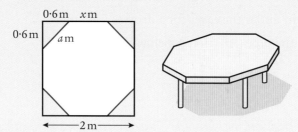

He cuts off the right-angled triangles shown, which have shorter sides of length 0·6 m.

a What is the value of x?

b Calculate the value of a.

c Is the octagon that he gets regular?

d Should the 0·6 m have been slightly more or slightly less for him to get what he wants?

4 A hiker is walking towards a crossroads.

He is a mile from it at A when he decides to take a shortcut.

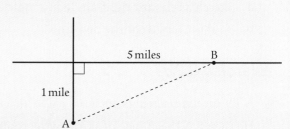

He cuts across country, aiming for a point B, 5 miles from the crossroads.

a What is the length of the shortcut?

b On the road, the hiker walks at a constant speed of 3 miles per hour.

On the rougher terrain of the shortcut he can only average 2·5 miles per hour.

How long will it take him to get from A to B:

i by road ii by the shortcut?

Comment.

c Suppose, instead, he had aimed for a point 4 miles from the crossroads.

Comment.

5 An engineer uses a photograph to help him estimate sizes on the Forth Bridge.

He places a picture in a computer document, then draws a line AB.

Double-clicking the line AB gets the computer to reveal that the distance from A to B is 7·1 cm horizontally and 1·7 cm vertically.

a What is the length of the line AB on the photograph?

b It is known that the height of the tower AC is actually 100·8 m.
On the photograph it is 3·6 cm.

What does 1 cm on the photograph represent in real life?

c What is the length on the actual bridge from A to B?

6 Investigate some suitable structure near your school using a digital camera, e.g. a statue:

- where not all of the relevant sizes are horizontal or vertical
- where at least one actual size is known (e.g. we can measure the height of a stone in the pedestal).

Use the drawing facilities of the computer to get horizontal and vertical lengths.

Use Pythagoras' theorem to get others.

Write a report on your findings.

8.3 Finding a shorter side

Example 1

A triangle is right-angled with a hypotenuse of size 45 m.

One of the shorter sides has a length of 27 m.

What is the length of the third side?

A sketch gives the situation:

Since the triangle is right-angled, we may use Pythagoras' theorem:

$$x^2 + 27^2 = 45^2$$
$$\Rightarrow x^2 = 45^2 - 27^2 = 1296$$
$$\Rightarrow x = \sqrt{1296} = 36$$

The third side is 36 m long.

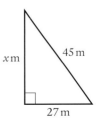

Example 2

The size of a TV screen, when quoted in an advert, is the length of the diagonal.

A 31-inch screen has a height of 15·2 inches.

How wide is the screen? Give your answer correct to 1 decimal place.

Making a sketch we see that one-half of the screen is a right-angled triangle.

The triangle is right-angled, so Pythagoras' theorem is appropriate.

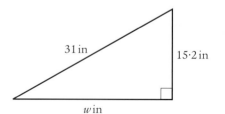

$$w^2 + 15{\cdot}2^2 = 31^2$$

$$\Rightarrow w^2 = 31^2 - 15{\cdot}2^2$$

$$= 729{\cdot}96$$

$$\Rightarrow w = \sqrt{729{\cdot}96}$$

$$= 27{\cdot}0177...$$

The width of the screen is 27·0 inches (to 1 d.p.).

Exercise 8.3A

1 Find the size of the third side in each triangle.

a

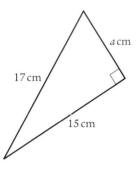

b

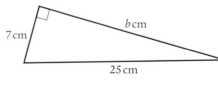

c

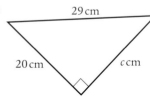

2 Calculate the exact value of the third side in each triangle.

a

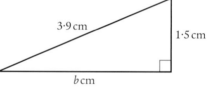

b

c

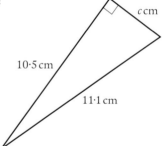

3 A naturalist is photographing seals as they bask on a dyke that runs into the sea.

This dyke is 34·8 m long.

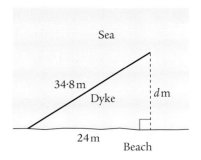

The photographer sets up his camera on the beach, 24 m from the end of the dyke, lined up with the other end of the dyke at sea.

He needs to set the distance to the end of the dyke in the camera to focus.

What is this distance?

4 Flying a kite, you can work out the length of cord that is played out. Suppose this is 54·4 m.

You can also find, with the help of a friend, the distance on the ground to the point immediately below the kite ... 25·6 m.

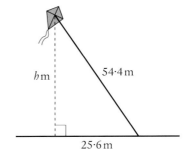

How high is the kite if we assume that, when the cord is taut, it forms a straight line?

5 An ornithologist tagged a redshank and released it.

It was picked up by GPS two days later 97·5 km away.

This point was 90 km east of its release point.

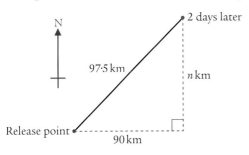

How far north of the release point is it?

6 a An isosceles triangle has equal sides, each of length 319 mm.
Its altitude is 220 mm.

Calculate the length of its base.

(Hint: the altitude will split the triangle into two right-angled triangles.)

b A kite has sides of 27 mm and 40 mm.

The diagonal, which is bisected by the axis of symmetry, has a length of 48 mm.

Calculate the length of the other diagonal.

Exercise 8.3B

1 A window cleaner has an adjustable ladder.

He uses a sound point on the ground 2 m from the wall to anchor the ladder.

The ladder can extend from 8 m to 16 m.

a How far up the wall will it reach if it is 8 m long?

b If it is extended to its full length, how far up the wall will it reach?

c The ladder is fully extended. By mistake the foot of the ladder slips to a point 4 m from the wall.

By how much does the top of the ladder slide down the wall?

2 'As the crow flies', Fort William is 73·9 miles from Glasgow.
It is 32·9 miles west of Glasgow.

a How far north of Glasgow is it?

b Asking for directions on the internet, using the Route Searcher software, gives the distance as 102 miles with a journey time of 3 h 42 min.

i Express the difference between the direct distance and the distance by road as a percentage of the direct distance.

ii Using a journey time of 4 hours as an estimate, what average speed does the website suggest for the journey?

What kind of road does this suggest: motorway or country?

3 The SS *Waverley* is 73 m long.

When it sails by at an angle it appears shorter (it is foreshortened).

Viewed from above we can see the situation:

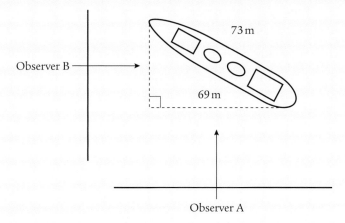

Observer A gets the impression that the ship is 69 m long.

a   What length does observer B see as the apparent length of the ship?
b   If both observers see the ship as the same length:
   i   what is the length they see
   ii in what direction is the ship sailing (A is facing due north)?

4 It is a fact that an angle drawn from the ends of a diameter to the circumference of a circle is a right angle.

A circle with centre O and diameter AB has a triangle ABC drawn in it, where C lies on the circumference.

The radius of the circle is 5 cm.

a   What is the area of the triangle when BC is:
   i   2 cm   ii   3 cm   iii   4 cm?
b   What length is BC when the area of the triangle is as large as possible?

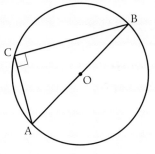

8.4 Point-to-point

Finding the distance between points on a grid is easy as long as the points are on the same vertical or horizontal line.

When they lie on a line that is at an angle to the axes then we can use Pythagoras' theorem.

Example 1

Find the distance between P(2, 1) and Q(8, 9).

Plot the points and form a right-angled triangle whose shorter sides are parallel to the axes.

From (2, 1) to (8, 9) is $8 - 2 = 6$ units in the x-direction.

From (2, 1) to (8, 9) is $9 - 1 = 8$ units in the y-direction.

The distance we wish to find is the hypotenuse of the triangle.

$$PQ^2 = 6^2 + 8^2$$
$$\Rightarrow PQ^2 = 36 + 64 = 100$$
$$\Rightarrow PQ = \sqrt{100} = 10$$

The distance between P and Q is 10 units.

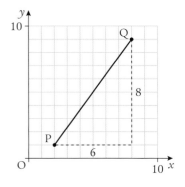

Example 2

Find a point in the first quadrant that is 13 units from the origin and has an x-coordinate of 5.

A quick sketch forms a right-angled triangle:

$$13^2 = 5^2 + h^2$$
$$\Rightarrow h^2 = 13^2 - 5^2$$
$$\Rightarrow h^2 = 169 - 25 = 144$$
$$\Rightarrow h = \sqrt{144} = 12$$

The required point has coordinates (5, 12).

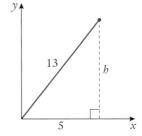

Exercise 8.4A

1 Find the distance of each point from the origin.

    a (15, 8)   **b** (12, 16)   **c** (48, −20)   **d** (−4, 3)   **e** (−20, 21)

2 Calculate the distance between each pair of points.

    a A(3, 4) and B(6, 8)   **b** C(2, 7) and D(14, 12)

    c E(5, 1) and F(13, 16)   **d** G(−1, 2) and H(23, 9)

3 By placing a grid on a photograph we can analyse an object.

The *Antigua* is a barquentine. Various parts of the rigging have been highlighted.

    a Calculate, to 1 decimal place, the length of:

        i TV   **ii** RS.

    b **i** Which is longer, RS or PQ?

        ii By how much?

4 Archaeologists are studying a stone circle.

Using its centre as the origin and suitable units to form a grid, they find stones at A(84, 13), B(36, −77), C(−80, 18) and D(−40, −75).

One of the stones they discover is not on the circle.

a What is the radius of the circle?

b **i** Which stone is not on the circle?

   ii How far from the centre is this stone?

5 **a** A right-angled triangle has shorter sides 48 cm and 36 cm.

   Calculate its perimeter.

b A second right-angled triangle has hypotenuse 65 cm and one side of 16 cm.

   Calculate its perimeter and comment.

c Does a right-angled triangle with shorter sides 40 cm and 42 cm share this property?

Exercise 8.4B

1 A triangle has vertices P(25, 60), Q(56, 33) and O(0, 0).

a Prove that the triangle is isosceles.

b Calculate the perimeter of the triangle.

2 Kate thought she'd drawn an equilateral triangle by joining the points P(3, 8), Q(9, −3) and R(−3, −3).

a Prove that she was wrong.

b What is the distance from R to the midpoint of PQ?

c What is the area of the triangle?

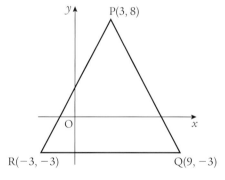

3 Three houses share a communal aerial.

The aerial will be placed in a position agreed by all and each householder will pay for the wiring to their own house.

Using a suitable grid and scales, the houses are located at points A(100, 15), B(80, 55) and C(24, 97).

a Who gets the best deal if the aerial is located at (−20, −10)?

b Show that the location (−20, −20) is the place where all pay the same.

c Calculate the sum of the three distances for both locations and comment.

4 **a** Find two points that are 17 units from the origin and have an *x*-coordinate of 8.

   b Find two points that are 61 units from the origin and have an *y*-coordinate of 11.

   c Find a point that is 29 units from the origin with an *x*-coordinate of 20 **and** is 25 units from the point (5, 1).

5 A rectangle has vertices A(3, 6), B(5, 2), C(−5, 2) and D.

   a State the coordinates of D.

   b Calculate, correct to 2 decimal places, the length of:
     i AB   **ii** AC.

   c Calculate the exact value of the area of the rectangle.

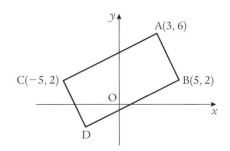

Preparation for assessment

1 Calculate the value of *x* in each of the triangles.

The diagrams are **not** drawn to scale. All distances are in centimetres.

a

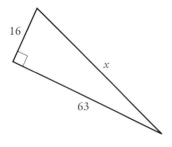

b

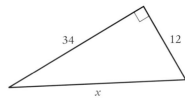

c

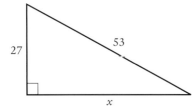

d

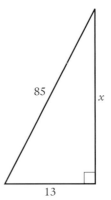

2 On the Clyde a tall ship is moored.

Michael wanted to know the length of the bowsprit.

In a document he placed a photo, drew a line along the bowsprit and double-clicked it to get its size. He was given the detail that its height was 4 cm and its width was 8·9 cm.

Use this information to find the length of the bowsprit in the photo.

3 The *Emerald Ice* sails from Ardniven to Barrcraig.

Ardniven is 26 km east of Barrcraig and 17 km north of it.

Calculate the distance between Ardniven and Barrcraig, correct to 1 decimal place.

4 The wings of a microlite are in the shape of a V-kite.

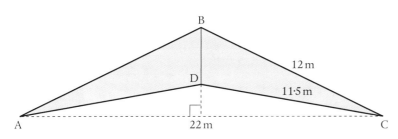

The wingspan AC is 22 m. The leading edge of the wing is 12 m and the trailing edge is 11·5 m.
How broad is the wing, BD?

5 A set of steps down to the river is used by a fishing club.
A handrail is required.
The diagram shows what would be needed for a five-step set.

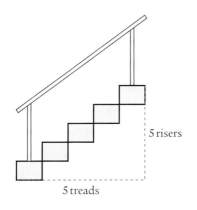

Each step is identical, having a tread of 35 cm and a riser of 40 cm.

There are 16 steps.

a From bottom to top:
  i how high do you climb
  ii how far forward have you gone?

b What length of handrail is required?

 6 Remember Margaret? She wanted to get some facts about the Ford Anglia 1960 model using old photos from the Museum of Transport.

All she knew was that the floor tiles were squares of side 310 mm (1 foot).

Is there enough information in the picture for her to estimate the size of the wheel base?

The wheel base is the distance between the two corresponding points on back and front wheels, for example, the centres or the points where the wheels touch the ground.

9 Budgeting and costing

Before we start...

Each year, the government calculates how much money will be gathered through taxes and other sources ... **income**.

It then has to plan how much it needs to spend ... **expenditure**.

In certain years, the expenditure is greater than the income.

Then the government has to borrow from international markets.

The Chancellor of the Exchequer tells the country his plans for government spending on Budget Day.

The two bar graphs show the government **expenditure** and **income** for a certain year.

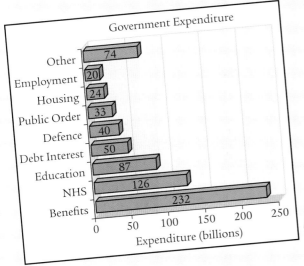

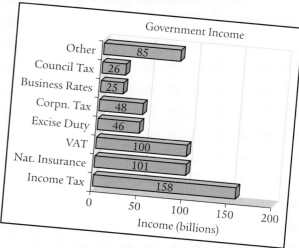

Do the figures indicate that the government will have a surplus or will they need to borrow?

How much will the surplus/borrowing requirement be?

What you need to know

1   Ronnie is saving to buy his mother a bottle of her favourite perfume for her birthday.

   He knows he can save £2·80 a week for the present.

   His mother's birthday is 15 weeks away.

   What is the most expensive bottle of perfume he could afford?

2 Anna earns £65 a week working part-time in a supermarket.

   She spends an average of £48·50 each week and saves the rest.

   How much did she save over a four-week period?

3 It costs Nareena £4·60 a day to travel by train to her work.

   Her take-home pay for working five days is £170.

   How much is left after paying her travel costs?

4 Over a seven-day period, Tom spent £8·76, £11·39, £7·47, £6·88, £9·15, £7·08 and £8·44.

   His weekly take-home pay was £235.

   How much did he have left to pay for rent, energy bills, food and other household expenses?

5 Mhairi managed to save £47 in a week where her total spending was £138.

   What was her total income for that week?

6 Ali takes home £64 for working part-time at the local petrol station.

   He intends to save a quarter of this.

   How much can he spend?

9.1 Budgeting and the home

Annual income twenty pounds, annual expenditure nineteen pounds nineteen and six, result happiness.

Annual income twenty pounds, annual expenditure twenty pounds nought and six, result misery.

Mr Micawber ... Charles Dickens (1850)

Budgeting is the process of balancing income and expenditure.

The aim is to ensure that the total income is greater than the total spending.

If this occurs, there will be a surplus that can be saved.

If the total spending is greater than the total income, there will be a deficit, which leads to debt.

Example 1

The Wilson family are budgeting for the month of September.

They set up spreadsheet tables and enter the income and expenditure.

These are the figures they arrived at:

<>	A	B	C	D
1	Income		Expenditure	
2	Salary Mr Wilson	£1150	Mortgage	£640
3	Salary Mrs Wilson	£1470	Utilities	£180
4	Child benefit	£120	Council Tax	£120
5			Food	£350
6			Petrol/travel	£280
7			Insurance	£55
8				

How much will the Wilson family have available to save or spend on other items?

Total income: £1150 + £1470 + £120 = £2740.

Total expenditure: £640 + £180 + £120 + £350 + £280 + £55 = £1625.

Difference: £2740 − £1625 = +£1115.

Amount available for saving or spending on other items is £1115.

Example 2

Donna's weekly take-home pay was £265 and her regular bills for the week totalled £215.

a   How much did she have each week for other spending?

b   Unfortunately, Donna is forced to work part-time due to company cut-backs.
Her weekly take-home pay is reduced to £165 while her bills remain the same.
Explain how this affects her budgeting.

c   Donna saw an advert for short-term loans from Rip-U-Off Loans.
She borrowed £200 to be paid back in four weeks to pay her rising debt.
The loan company expected the £200 plus 80% interest to be repaid at the end of the four weeks.
By considering her regular weekly bills, her loan repayment and the interest, how much will Donna be in debt four weeks after taking out the loan?

a   Surplus for other spending = £265 − £215 = £50.

b   £165 − £215 = −£50
She is now building debt at a rate of £50 a week

c   Interest on loan = 80% of £200 = £160.
After four weeks:
regular debt = 4 × £50 = £200, paid off by loan of £200 ... balance £0
repayment of loan = £200 + £160 = £360.
If she hadn't taken the loan, she'd be £200 in debt.
Taking out the loan, she has £360 of debt. She's a lot worse off.

Exercise 9.1A

1 Copy and complete the table.

The first column is done for you.

	a	b	c	d	e	f	g
Income	£245	£336	£241	£187·54		£427·09	£349·28
Expenditure	£188	£297	£307		£452·63		£349·28
Surplus	£57		–	£32·86			
Shortfall	–				£132·89	£169·77	

2 Andy is a shepherd. His weekly income is £345.

He lists his weekly expenditure:

Rent and council tax	£135·70
Travel	£35·65
Food	£63·82
Sports clubs	£18·75
TV, phone and internet	£28·67
Heating and lighting	£32·80

How much is left for saving or other spending?

3 This was the monthly expenditure for the Murphy family three years ago.

At the time their monthly income was £2600.

Mortgage	£685·00
Council tax	£135·80
Travel	£47·50
Food	£325·00
Entertainment	£125·00
Utilities	£145·70
Household items	£143·00

Since then, their expenditure has risen by 10%.

The monthly family income has fallen by 5%.

a What was their monthly surplus three years ago?

b What is their monthly surplus now?

c Discuss how this sort of trend leads to a recession.

4 To begin with, the Adams family had savings of £1250.

In the first week after that, they found themselves short by £72 and they got it from their savings.

In the second week, they found they had a surplus of £40, which they put into savings.

The graph shows how, over the next ten-week period, their weekly budget changed.

What is their new savings balance after the ten weeks?

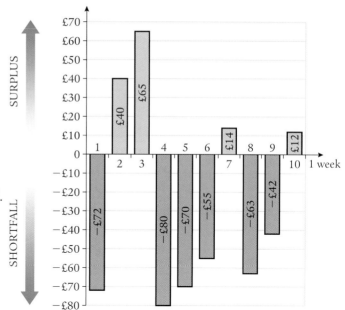

5 Elaine is a student who earns £65 a week for working part-time.

The table shows the normal breakdown of her expenditure for a week.

Magazines	£4·80
Music downloads	£6·60
Cosmetics	£12·80
Mobile phone	£10·70
Socialising/cinema/concerts	£15·00
Bus fares	£15·10

She wants the new model of mobile phone but this will cost £13·30 a week.

To budget for this, she reduces one of her expenditure amounts by 10% and another by 20%.

Which two items of expenditure did she choose to reduce?

Exercise 9.1B

1 Brian has a weekly take-home pay of £315.

The bar graph shows his average weekly expenditure.

He needs to save more to pay the deposit on a motor bike.

He decides he will cut all items of expenditure by 10%, except his rent which is fixed.

How much will he save each week when he 'tightens his budget'?

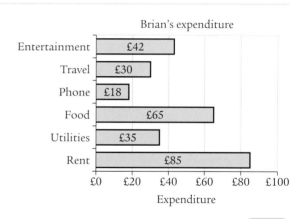

2 Louise has a monthly take-home salary of £1350.

The table shows her average monthly expenditure.

She has six months to save £858 that will pay for a holiday she is taking.

a How much short of her target will she be?

b For the next six months, she decides to reduce her expenditure on 'Entertainment' and 'Other spending' by x%.

What is the minimum value of x that will allow her to save enough for the holiday?

Rent/council tax	£475
Utilities	£60
Food	£235
Entertainment	£200
Other spending	£260

3 The pie charts show the monthly income and expenditure for the Donaldson family.

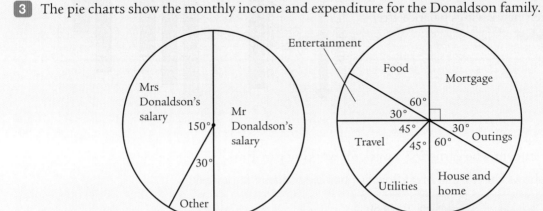

Monthly income   Monthly expenditure

Mrs Donaldson's monthly income is £1500.

a What is the total monthly income for the Donaldson family?

b The monthly mortgage repayment for the Donaldsons is £730.

How much is their total monthly expenditure?

c All surplus money goes into savings.

How much do the Donaldsons save each year?

9.2 Budgeting and holidays

 ## Class discussion

When planning a holiday abroad, there are certain things you need to consider.

- What is the most effective way to turn your money into foreign currency?
- How does the time/date of flights affect the cost of the holiday?
- Thinking of expense, which items should you buy here and which ones abroad?
- What is the cost of taking luggage on holiday?
- What type of insurance should you have?

Example 1

Jamie and his three friends plan to spend two weeks in the Greek islands.

They have agreed the accommodation but are comparing travel costs.

Athens Air return flight £235 per person			*Trojan Travel* return flight £199 per person	
Flight out	Departs 11 25		Flight out	Departs 22 40
Flight out	Arrives 17 40		Flight out	Arrives 05 00
Return flight	Departs 13 10		Return flight	Departs 23 45
Return flight	Arrives 15 35		Return flight	Arrives 02 10
One piece of hand luggage up to a maximum of 10 kg is free			*One piece* of hand luggage up to a maximum of 10 kg is free	
Other luggage up to 15 kg costs £18 per bag, single flight			Other luggage up to 22 kg costs £22 per bag, single flight	

Here are their alternatives.

What are their budgeting options?

Athens Air:

- Four flight tickets at £235 = £940.
- Good flight times so they can take public transport between the airport and accommodation. So by public transport: four return tickets at £5 = £20 (estimate).
- If they split clothing among three cases (as well as the hand luggage). Cost for luggage = 3 × £18 = £54 (one way). Total baggage cost = £108.
- Total cost for four people = £940 + £20 + £108 = £1068.

Trojan Travel:

- Four flight tickets at £199 = £796.
- Awkward flight times so they need to take a taxi between the airport and accommodation. Two journeys at £25 = £50.
- If they split clothing among two cases (as well as the hand luggage). Cost for luggage = 2 × £22 = £44 (one way). Total cost for baggage = £88.
- Total cost for four people = £796 + £50 + £88 = £934.

They can make Trojan Travel 'work' for them, but with a bit of inconvenience.

Their decision: is it worth it?

Example 2

Louise has saved £600 to spend on a holiday in Italy.

She wants to change her money into euros.

Option 1: She can convert all her money in the UK at a rate of €1·18 per pound.

Option 2: If she's not happy carrying a lot of money with her, she can use her bank card. She gets a better rate, €1·22 per pound, but is charged £4·50 each time she uses her card. She estimates she would use her card four times during the holiday.

Assuming she doesn't want to spend more than £600, which of the options gets her more euros?

169

Option 1:

  $600 \times €1.18 = €708.$

For her £600, Louise gets €708.

Option 2:

She uses the card four times: $4 \times £4.50 = £18.$

So she has $£600 - £18 = £582$ left to exchange.

  $582 \times €1.22 = €710.04.$

For her £600, Louise gets €710.04.

Option 2 provides better value for money.

 What are the benefits and dangers of each option?

Exercise 9.2A

1   The Dunlop family are budgeting for their next holiday.

The table gives a breakdown of costs.

Flights	£920
Accommodation	£635
Transfers	£40
Insurance	£65
Spending money	£700

They intend saving for the next 40 weeks to cover all the costs.

How much should be set aside each week to cover the full cost?

2   a   Lorna saved £500 to spend during a ten-day holiday abroad.

She exchanged her money when the rate was €1.17 per pound.

What was the average number of euros she could spend each day?

   b   Her friend also saved £500 but exchanged her money abroad.

She used an ATM four times with a charge of £4.50 each time. These charges came out of the £500 she had saved. She received an exchange rate of €1.21 euros per pound.

What was the average number of euros she could spend each day?

3   The Hussain family plan to save £80 each week to cover the cost of a holiday.

They go on the same holiday each year and save for 35 weeks before the holiday.

   a   How much have they budgeted the holiday to cost?

   b   They are told that the holiday will be 10% more expensive than they think.
       What will the holiday cost?

   c   How much will they now need to save for each of the 35 weeks?

   d   What do you notice about this new weekly savings figure?

4 Jenna did a web search for the best company from whom to buy travel insurance.
She chose 'Scottish Sun Insurance' who gave the following information on annual charges.

One trip to EU country	£37
Multi-trips to EU countries	£63
One trip to non-EU country	£48
Multi-trips to non-EU countries	£79
Involvement with extreme sports	£15 per trip

   a Jenna intends to take a trip to Crete to do wind surfing and later a trip to Cyprus to do some microlite flying. Both these activities are classified as extreme sports.
How much will she need to budget to cover her insurance premium for the year?

   b Her friend Anna, is going with her to Crete, to wind surf, but taking a separate trip to Rio de Janeiro for a sunbathing holiday.
How much more expensive is her holiday insurance?

 c Discuss the dangers of going on holiday without travel insurance.

Exercise 9.2B

1 Mr and Mrs Gavin plan a two-week holiday to the Vendée region of France.
They can fly and pay for accommodation and car rental when the get there or they can drive with their caravan on tow, taking the ferry to get to the Continent.
The tables show the costs they have budgeted for.

Fly and rent	
Return flights	£620
Car hire in Vendée	£275
Accommodation	£960

Caravanning	
Ferry charges (return)	£128
Caravan site charges	£345
Cost of petrol per litre	130p

   a What is the total cost of the holiday when flying?

   b They have calculated that the return trip from Scotland to France will be 2800 km.
They expect to travel 10 km on a litre of petrol.
How many litres of petrol will they need?

   c What is the total cost for petrol?

   d What is the total cost of the caravan holiday?

2 Sunshine Travel offer a Mediterranean holiday for £420 per person.
To encourage larger groups they offer the following savings:
- second person in a group gets a 10% discount
- third person in a group gets a 20% discount
- fourth person in a group gets a 30% discount.

   a What is the total cost of the holiday for a party of four?

   b Tantastic Travel charge £400 per person but the fourth person in a group only pays half price.
How much would the party of four save by travelling with Tantastic Travel instead of Sunshine Travel?

3 The Watson family have budgeted to save £67·50 a week, for 40 weeks, to pay for their holiday.

The all-inclusive luxury holiday they want normally costs £4500.

The travel agent is prepared to discount the holiday by 30%.

a How much short of the discounted price will their total savings be?

b What is the minimum percentage discount that will allow their savings to cover the cost of the holiday?

c The travel agent is prepared to discount the holiday by a third.

How much **extra** each week would they need to budget for to ensure their savings covered the cost of the holiday when discounted by a third?

9.3 Budgeting and cars

Most young people cannot wait for their seventeenth birthday, so that they can start taking driving lessons.

Having a car or a motorbike gives you independence and opens up many travel and job opportunities.

It also carries responsibilities, like staying within speed limits and driving with care and attention, and many costs, such as insurance, road tax, a valid MOT certificate ... and a car!

When buying a car, it is important to budget for the other expenses that go along with owning a car.

Example 1

Bryan had saved £800 and wanted a car as soon as he obtained his driving licence.

He bought a ten-year-old car for £550, thinking he could afford to run it.

The car was due for an MOT in four months and had road tax that expired in three months.

What budgeting factors should he have considered as well as the cost of the car?

It is illegal to drive without insurance, a valid MOT certificate, and proof of having paid road tax.

The costs of these vary according to various factors, e.g. age of driver, age of car, etc.

- Insurance ...   £760   (needed immediately)
- Road tax (6 months) ...   £66   (needed in 3 months)
- MOT certificate ...   £35   (needed in 4 months)
- Typical bill for repairs after MOT ... £150   (needed in 4 months).

Total cost of car after four months = £550 + £760 + £66 + £35 + £150 = £1561.

Shortfall = £1561 − £800 = £761.

Bryan will need to find another £761 over the next few months to cover car costs.

Example 2

Laura wants to buy a secondhand car from a car dealer, costing £4800.
She has saved £1200 but needs to finance the remainder.

Her options are:

- a bank loan for three years ... repayments £118 a month
- the car dealer's hire purchase deal ... four years at £94 a month.

a How much does the bank loan cost?

b How much more expensive is the car dealer option?

c Why might Laura opt for the more expensive option?

a Total repaid to bank = £118 × 36 = £4248.
Bank charge = £4248 − £3600 = £648.
Borrowing £3600 for three years costs £648.

b Total repaid under car dealer's deal = £94 × 48 = £4512.
Car dealer's option is more expensive by £4512 − £4248 = £264.

c Laura may choose the more expensive option because, although she would need to pay twelve more instalments, her monthly budget would be £24 less for the car.

Exercise 9.3A

1 A secondhand car, costing £3500, can be bought for a 20% deposit and then 24 monthly payments to cover the remainder of the cost.

   a What is the size of the deposit?

   b What is the size of a monthly repayment (rounded to the nearest necessary penny)?

   c The first 23 payments are made at the rounded rate.

   The last payment will be less because of the rounding. What size will the final payment be?

2 A motorbike costing £1800 can be bought for a 10% deposit and 12 monthly payments to cover the outstanding amount.

   Tina has saved £200 and has budgeted for repayments of £120 a month.

   a Does she have enough for the deposit?

   b Has she estimated the monthly repayments correctly?

3 Ted wanted to buy a replacement works van for his business.

   The £4500 van he wants can be bought for a 20% deposit and 24 monthly payments, to cover the remainder.

   a How much is the deposit?

   b How much is a monthly repayment?

   c Before buying the van he sold his old van for £1700. He has budgeted for monthly payments of £120.

   Do the calculations needed to show that he can afford the new van.

4 A motoring organisation did a survey to study the relation between 'age of car' and 'average repair costs' needed to gain an MOT certificate.

They found that the average cost of repairs for a three-year old car was £40.

The cost of repairs for subsequent years was 50% higher than the previous year's figure.

Angela wants to save for her next MOT test, when her car will be six years' old.

She has to pay £38 for the test and any repair bill the tester deems necessary.

a Estimate how much she will need to save each week if the MOT is due in 15 weeks.

b What is the significance of starting the survey on three-year-old and not one-year or two-year-old cars?

5 Maureen bought a new car and was told her tyres could need replacing after 30 000 km.

She drives an average of 20 000 km a year.

a After how many months should she expect to replace her tyres?

b The cost to replace a set of tyres is estimated to be £225.

How much should she save each month to pay for the replacement tyres?

c Investigate:
  - the conditions that make a tyre illegal
  - the penalties for driving with an illegal tyre
  - what affects the life expectancy of a new tyre.

Exercise 9.3B

1 Every car has a trade-in value. This is what a dealer is willing to give you for your old car when you buy a new one.

Misha's car presently has a trade-in value of £4000.

In the first year after this, the value will drop by 20%. Its value at the end of the second year will be 15% less than its value at the end of the first year.

She hopes to trade in her car in two years' time and buy a replacement costing £5600.

a Estimate what the trade-in value of her car will be then.

b How much extra will she need to save to cover the cost of the replacement car?

c She decides to save monthly for two years to make up the rest of the money needed.

How much will she need to save each month?

2 Alan was an erratic driver. He used his accelerator and brake inefficiently.

It was pointed out to him that if he was prepared to drive more carefully his petrol consumption could be reduce by 15%.

He normally used 80 litres of petrol a month and paid 130 pence per litre.

How much would he save in a year if he improved his driving technique?

3 Moira needs to replace **all four tyres** on her car.

She considered two types of replacement.

Tyre A			Tyre B	
Cost per tyre	£42		Cost per tyre	£55
Expected milage	20 000 km		Expected milage	30 000 km
Replacement valves	£3·80		Replacement valves	£3·80
Wheel balancing	£4·20		Wheel balancing	£4·20

  a What is the total cost of replacing the tyres with four tyres of type A?

  b What is the total cost of replacing the tyres with four tyres of type B?

  c Show your working to demonstrate why tyre B is the better option in the long term.

9.4 Budgeting and home insurance

Class discussion

What is insurance?

Why is it important?

Which kind of insurance is required legally?

What unusual things can be insured?

Example 1

Kerry contacted a number of insurance companies about insuring her house.

Each one of them asked the same questions.

1. How many rooms are in the house?
2. What percentage of your roof is 'flat roof'?
3. What is your postcode?
4. Is there a history of settlement in your area?
5. Are there trees close to your property?
6. Do you have an alarm and safety locks?
7. Is there history of flooding in your area?

Why would these factors affect the cost of insurance?

1. The cost of rebuilding a house, after serious damage, depends on the number of rooms.
2. Flat roofs are more likely to allow water penetration that causes damage.
3. Houses in certain areas (postcodes) are more likely to be burgled than houses in other areas.

4. Settlement causes major damage to houses and is expensive to repair.

5. Trees close to houses can cause serious damage when blown down in high winds.

6. Alarms and safety locks deter burglars and prevent break-ins.

7. The cost of repairing houses that have been flooded can be excessive.

Example 2

The annual cost of insurance for a certain type of house is £180 (called the premium).

The insurance company added a surcharge of 40% where there was a flat roof or the possibility of subsidence.

They also gave a discount of 15% for the security features within the house and the postcode area.

These percentages are calculated on the original premium.

a   How much was the surcharge added?

b   How much discount was given?

c   What was the actual insurance premium for the year?

a   Surcharge = 40% of £180 = £72.

b   Discount = 15% of £180 = £27.

c   Annual insurance premium = £180 + £72 − £27 = £225.

Exercise 9.4A

1   The *B-Safe Insurance Company* offered annual home insurance at £210.

They also offered a 10% discount on this figure if you included contents insurance.

The content insurance premium was £65 for the year.

a   How much is the discount worth?

b   What was the total annual premium for house and contents?

2   A home-owner was quoted £240 for annual home insurance.

He was told that if he fitted an alarm to the house there would be a 15% discount.

a   What would he save on insurance by fitting the alarm?

b   What would be the discounted insurance premium?

3   Two houses of identical build were in the same city.

One had a postcode that meant the home insurance had a surcharge of 10%.
The other had a postcode that carried a 5% discount on home insurance.

The standard premium for insuring these houses was £260.

a   How much was the premium with the surcharge?

b   How much was the discounted premium?

c   What was the difference between the two premiums?

4 A small shop, in an area with a history of flooding, paid an insurance premium of £1260 last year.
Due to recent serious flooding, the premium this year will increase by 80%.

The shopkeeper budgeted to pay the insurance monthly.

a What were the monthly payments last year?

b What are the increased payments this year?

c What is the most obvious way of recouping the increased insurance costs?

Exercise 9.4B

1 A house is quoted home insurance at £280 for the year. The insurance company discover there is a large tree in front of the house. If this were to fall, it would do serious damage to the house. Because of this increase in risk, the insurance premium is increased by 75%.

a What would the increase to the premium be each year?

b The local tree-feller offered to remove the tree for a cost of £850. What advice would you give the home owner about the tree?

2 The table shows the annual cost of insurance quoted by three different companies.
They all offered a discount to encourage homeowners.

	Company A		Company B		Company C
Building	£195		£210		£175
Contents	£65		£58		£52
Discount	2·5% for every year without a claim		15% after 4 years without a claim		10% for buying both house and contents insurance

What would each company charge a homeowner who had not made an insurance claim for six years?

3 An insurance broker worked on the basis that the cost of insuring contents would be 30% of the size of the building insurance cost.

a What would it cost to insure contents, when the building insurance was £170?

b What would the building insurance be if the contents cost £45 to insure?

c What would the building insurance cost if the insurances both together cost £260?

9.5 Budgeting and car insurance

 Class discussion

In car/motorbike insurance, why do the following affect the annual premium?

- Age of driver.
- Size of car engine.
- Number of years without a claim.

Example 1

Anne is over 45 and has been driving for 12 years without an accident.
She wants to insure a high-performance sports car.
Her basic insurance premium would be £340.
There is a 40% surcharge added because of the type of car.

a What would Anne's actual insurance premium be for the year?

b Her son is a newly qualified driver who wants to insure the same car.
His basic premium will be four times his mother's basic premium. There will also be an 80% surcharge because of the type of car.
What would his actual insurance premium be for the year?

a Surcharge = 40% of £340 = £136.
Anne's insurance premium for year = £340 + £136 = £476.

b Son's basic premium = 4 × £340 = £1360.
Surcharge = 80% of £1360 = £1088.
Son's insurance premium for year = £1360 + £1088 = £2448.

Exercise 9.5A

1 Tom's basic car insurance premium would be £480.
He is entitled to a reduction of 45% because of his age and no-claims discount.
What would his actual insurance premium be for the year?

2 Olivia is 21 years' old and drives a popular small car.
The normal insurance quote would be £380.
Due to her age, this figure is doubled.
She is entitled to a 20% no-claims discount on the increased premium.
What is her actual insurance premium for the year?

 3 **a** Investigate what is meant by a no-claims discount (NCD).

b Tom should pay £620 for his car insurance but has built up a NCD of 60%.
His friend Iain has to pay car insurance of £585 in full, as he lost his NCD, after causing a major accident through carelessness.
How much more does Iain pay than Tom?

4 A 20-year-old motorcycle fanatic wants to buy a bike costing £3100.

The cost of insurance for one year will be three-quarters the value of the bike.

How much will she need to pay for the bike and insurance combined?

Exercise 9.5B

1 Linda's no-claims discount rose from 20% last year to 30% this year.

Last year her premium without discount was £560.

This year her premium without discount has risen to £620.

How much has her actual premium reduced this year compared to last year?

2 Halifax Insurance gives discounts for multiple insurance policies.

Mr and Mrs Black insure both cars and their house with Halifax.

	Mr Black	Mrs Black	House
Premium	£680	£520	£340
Discount	60%	40%	30%

The table shows premiums before discount and the percentage discount.

What is the total insurance paid to Halifax Insurance for the year?

3 The Watt family want to insure their car and house.

They compare two different companies for cost.

The table shows the two quotes they received.

	Auto & Home Co.			Blanket Cover Co.	
	Car	Home		Car	Home
Premium	£645	£360		£545	£435
Discount	60%	30%		50%	40%

a Which company will be cheaper for the car and home package?

b What is the difference between the two companies?

Preparation for assessment

1 The weekly income for a family is £465.

Their weekly expenditure is £428.

How much would they be able to save in a year?

2 Two years ago, the Lawson family had an average monthly income of £2300.

Their average monthly expenditure was £1950.

  a What was the maximum amount they could have saved in that year?

    Last year their monthly income had reduced by 3% and their expenditure had risen by 5%.

  b How did this affect the maximum possible savings, compared to the previous year?

3 a Tom exchanged £500 to euros, in a Bureau de Change, at a rate of €1·18 per pound.

    How many euros did he get?

  b Pam took her ATM card with her on holiday and got money from the machines at a rate of €1·22 per pound.

    She used the card four times and was charged £3·50 each time she used her card.

    The total amount she had budgeted for was £500, which had to cover charges as well.

    How many pounds were available to exchange into euros?

  c How many euros did she have, to spend on holiday?

4 The table shows the budgeting the Hussein family did when planning for their summer holiday.

Flights	£870
Accommodation	£730
Transfers	£65
Insurance	£75
Spending money	£960

  a What was the total cost of the holiday?

  b They have 40 weeks to save the money needed.
    How much will they need to save each week?

5 A motorist sees a car in the showroom valued at £8500.

He is told his present car is worth £3100.

The car dealer says he can pay the remainder over three years by monthly repayments.

What is the monthly repayment, assuming no interest is added?

6 Denise is comparing different ways of paying for a new fitted kitchen costing £4800.

  • She can take out a bank loan over three years with total interest of 12% added.

  • She can opt for monthly repayments of £120 over a four-year period.

  a How much interest would she repay to the bank?

  b What would her monthly repayment be on the bank loan?

  c What is the total cost, using the four-year option?

  d Why might Denise opt for the four-year option?

7 The Chan family was quoted premiums of £380 and £160 for house insurance and contents insurance, respectively.

The company also offered discounts of 30% for the house insurance and 25% for the contents insurance.

What would the total premium be after discounts?

8 Brian's no-claims discount rises from 20% to 30%.

His car insurance premium, before discount, rose from £480 to £510.

By how much will his actual insurance rise or fall?

9 Remember how, each year, the government calculates how much money will be gathered through taxes and other sources (**income**)? It then has to plan how much it needs to spend (**expenditure**).

In certain years, the expenditure is greater than the income. Then the government has to borrow from international markets.

The Chancellor of the Exchequer tells the country his plans for government spending on Budget Day.

The two bar graphs show the government **expenditure** and **income** for a certain year.

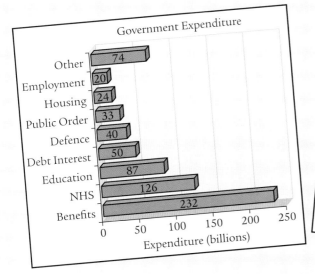

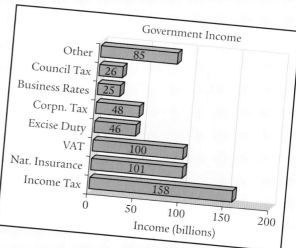

Do the figures indicate that the government will have a surplus or will they need to borrow?

How much will the surplus/borrowing requirement be?

10 Scale drawing

⏸ Before we start...

At the end of the chapter you are asked to consider the problem of designing a kitchen.

What you learn during this chapter can be applied to the task.

Kitchen units come in standard sizes. Generally the space for the kitchen does not.

There are regulations about how close a power socket can be to a sink, about how the space should be ventilated, etc.

This information can often be found on the internet.

There are practical considerations: Why would you place the sink close to the window? Have you heard of the Work Triangle? What's to be bought? What will it cost?

There is a lot to be said for letting thoughts and plans develop over time.
So as you progress through the chapter, keep the task in mind.

Scale drawings can show lengths and distances that are too great to draw full-size.

You will need to make a scale drawing of the kitchen to help you plan exactly where surfaces, sinks, electrical points, units and appliances are to be positioned.

It's easy to change your mind on a plan – not so easy on the real thing.

What you need to know

1 Convert each length to centimetres:

    a 4 m   **b** 6·4 m   **c** 18 m   **d** 10 000 m   **e** 1 km   **f** 2·5 km.

2 A triangular plot of land is surveyed.
A scale drawing of the plot is shown.
Measure each side in centimetres.

Given that 1 cm represents 5 m,
write down the length of each side in metres.

3 A yacht leaves the marina, M, and sails 6 km south to Fairhaven, F.

It then sails 8 km west to Yairmouth, Y.

    a Using a scale of 1 cm to 1 km, make a scale drawing of its voyage.

    b Measure MY.

    How many kilometres is Yairmouth from the Marina direct?

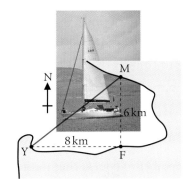

4 Calculate the values of:

    a i 6×20 **ii** 91×20 **iii** $10{\cdot}5 \times 20$

    b i 7×50 **ii** 46×50 **iii** $8{\cdot}4 \times 50$

    c i $400 \div 50$ **ii** $250 \div 50$ **iii** $95 \div 50$

    d i $450 \div 30$ **ii** $276 \div 30$ **iii** $843 \div 30$.

5 Five planes are on a holding path around Edinburgh Airport.

Calculate the three-figure bearing of each plane from the tower.

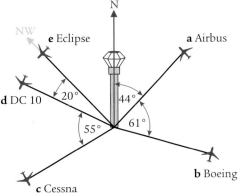

10.1 Scales

Designers, architects, engineers, builders, dressmakers and many others construct and use scale drawings as an important part of their work.

It is important that a scale is chosen to make the best use of the material on which it is drawn and to suit the purpose best: the bigger the scale, the more accurate is any answer obtained from the scale drawing.

The scale is always marked on the plan. It can be written in one of two ways:

For example, a scale of '1 cm represents 1 m' is equivalent to '1 : 100' with no mention of units.

The second way is often called the **representative fraction**: 1 unit on the drawing represents 100 units on the 'ground', whatever the units are.

A map maker is making something that is a **reduction** of the real thing.

Scientists often want **enlarged** models of atoms, insects, flowers, etc.

A scale of 25 : 1 tells us that 25 units on the drawing represent 1 unit on the 'ground'.

The amount by which an object is enlarged is called the **enlargement factor**.

The amount by which an object is reduced in size is known as the **reduction factor**.

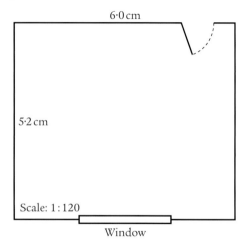

The flea

Scale: 25 : 1

We multiply by the factor to get the 'job done'.

If a map is drawn on a scale of 1 : 100,

the enlargement factor = 100   Map $\times$ 100 = ground

the reduction factor = $\frac{1}{100}$ Ground $\times \frac{1}{100}$ = map

Remember of course that multiplying by $\frac{1}{100}$ can be done by dividing by 100.

Example 1

On a plan the dimensions of a rectangular room measure 5·2 cm by 6·0 cm.
The scale of the plan is given as 1 : 120.

a What are the actual dimensions of the room in metres?

b What is the floor area of the room?

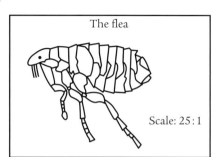

6·0 cm

5·2 cm

Scale: 1 : 120

Window

a Scale is 1 : 120.

$\Rightarrow$ 1 cm represents 120 cm.

$\Rightarrow$ 6 cm represents (120 $\times$ 6) cm = 720 cm = 7·2 m.

$\Rightarrow$ 5·2 cm represents (120 $\times$ 5·2) cm = 624 cm = 6·24 m.

The actual dimensions of the room are 7·2 m by 6·24 m.

b   Area = length $\times$ breadth = 7·2 $\times$ 6·24 = 44·9 m² (to 3 s.f.).

Note: the scale 1 : 120 in this example is the same as an enlargement factor of 120.

Example 2

Sam's garage is 3 m wide and 8.5 m long.

a What will be the dimensions of the garage on a plan with a scale of:
  i 1 : 10   **ii** 1 : 50   **iii** 1 : 100?

b You wish to draw the plan on a sheet of A4 paper.
Which of the above scales would make the best use of the paper?   *A sheet of A4 paper is 297 mm by 210 mm.*

a **i** A scale of 1 : 10 means a reduction factor of $\frac{1}{10}$.

$\Rightarrow$ Width on plan $= \frac{1}{10} \times 3$ m $= 0.3$ m $= 30$ cm.

$\Rightarrow$ Length on plan $= \frac{1}{10} \times 8.5$ m $= 0.85$ m $= 85$ cm.

So the dimensions on the plan are 30 cm by 85 cm.

ii A scale of 1 : 50 means a reduction factor of $\frac{1}{50}$.

$\Rightarrow$ Width on plan $= \frac{1}{50} \times 3$ m $= 0.06$ m $= 6$ cm.

$\Rightarrow$ Length on plan $= \frac{1}{50} \times 8.5$ m $= 0.17$ m $= 17$ cm.

So the dimensions on the plan are 6 cm by 17 cm.

iii A scale of 1 : 100 means a reduction factor of $\frac{1}{100}$.

$\Rightarrow$ Width on plan $= \frac{1}{100} \times 3$ m $= 0.03$ m $= 3$ cm.

$\Rightarrow$ Length on plan $= \frac{1}{100} \times 8.5$ m $= 0.085$ m $= 8.5$ cm.

So the dimensions on the plan are 3 cm by 8.5 cm.

b A sheet of A4 paper is 29.7 cm by 21 cm.

Scale of 1 : 10 ... the plan is far too large.

Scale of 1 : 100 ... the plan does fit A4.

Scale of 1 : 50 ... the plan fits A4 and is larger than the 1 : 100 plan.

So the plan with a scale of 1 : 50 is the best if we wish to take measurements from it.

Example 3

Mary makes a rough sketch of her bedroom and measures the lengths of the walls.

Draw an accurate plan of Mary's bedroom using a scale of 1 : 50.

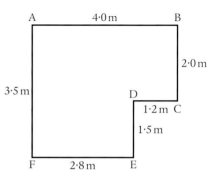

1 cm on the plan represents 50 cm of actual length.

Reducing each length using the factor $\frac{1}{50}$ ...

Wall	Length (cm)	Scaled length = length ÷ 50 (cm)
AB	400	8.0
BC	200	4.0
CD	120	2.4
DE	150	3.0
EF	280	5.6
FA	350	7.0

So the actual scale drawing looks like this:

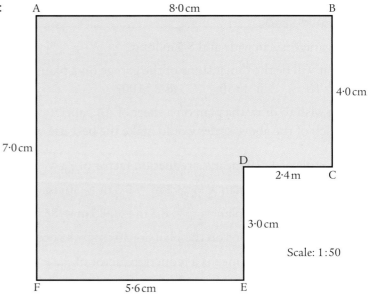

A — 8·0 cm — B

4·0 cm

7·0 cm

D

2·4 m — C

3·0 cm

Scale: 1:50

F — 5·6 cm — E

Exercise 10.1A

1 On a plan the dimensions of a rectangular room are 4 cm by 3 cm.
What are the actual dimensions of the room in metres if the
scale on the plan is:

a 1 : 100   **b** 1 : 150   **c** 1 : 80?

2 **a** Measure the height of the tree in centimetres.

b The scale is 1 cm represents 1 metre.
What is the height of the actual tree?

3 The pictures of the Eiffel Tower and the London Eye are to scale.

Scale: 1:5000

Full
height

Diameter

Scale: 1:3000

a Measure the height of the Eiffel Tower in centimetres and work out its actual height.

b **i** Measure the height in the picture of the London Eye in centimetres and calculate its actual height.

    ii Measure the diameter of the wheel in the picture and calculate the diameter of the actual wheel.

4 The diagram is a plan of Jack and Tessa's back garden.

  a Measure the length and breadth of the vegetable patch on the plan.

  b Use the scale to find the length and breadth of the actual vegetable patch.

  c What are the dimensions of the actual flower bed?

  d Find the diameter of the pond.

  e Jack and Tessa need to replace the panels of the fence that surrounds the garden.

    Each two-metre panel costs £9·50.

    i How many two-metre panels are required?

    ii What will the total cost be?

Scale: 1 cm to 2 m

Shed • Flower bed • Vegetable patch • Pond • Lawn • Patio

5 A scale model of a car has the following dimensions: length = 15 cm, breadth = 6·5 cm, height = 5 cm.

The scale is 1 : 28.

Calculate the actual dimensions of the car in metres.

6 The plan shows a plot of land with a house and double garage.

Use the scale to calculate, in metres:

  a the perimeter of the plot

  b the length and breadth of the garage

  c the length and breadth of the house

  d the area of the house, in square metres.

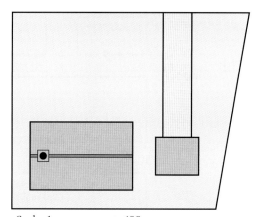

Scale: 1 cm represents 400 cm

7 Mrs Robertson's classroom floor is rectangular. It measures 12 metres by 8·6 metres.

She makes a scale drawing of the room, letting 1 cm represent 1 m.

  a What is the length of the classroom on the scale drawing?

  b What is its breadth?

  c What would its length and breadth be on the scale drawing if she used a scale of:

    i 1 cm : 2 m   **ii** 1 : 50?

  d What is the reduction factor when the scale is 1 cm : 2 m?

8 In a woodwork class, the students are making a nest of tables: three tables of different sizes that fit together.

The smallest table has a length of 405 mm, a breadth of 320 mm and a height of 375 mm.

The enlargement factor is 1·1 from one size to the next.

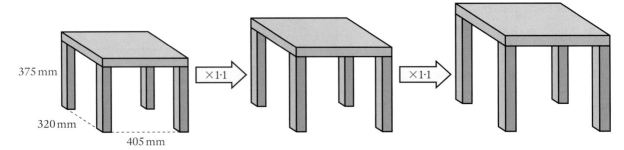

Find the length, breadth and height of the two larger tables.
(Give your answers to the nearest millimetre.)

9 The diagram below shows an extract from the street plan of the town of Whiteness.

The red dots represent the entrances to various places. Distances should be measured from dot to dot.

a Measure the distance on the plan from Alec's house to Ian's house.

b What is the actual distance in metres between the two houses?

c In the same way find the actual distances between:
  i Alec's house and the leisure centre
  ii Ian's house and the cinema (the shortest distance)
  iii Alec's house and the supermarket (the shortest distance).

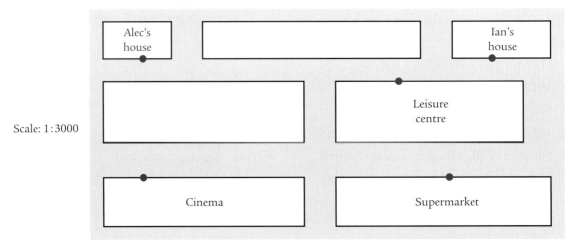

10 Sanji is opening a sandwich bar in the High Street.

The measurements of the premises are shown in the sketch.

a Using a scale of 1 cm to 2 m, make a plan of the sandwich bar.

b Write down the reduction factor.

c What are the dimensions on the plan of:
  i the kitchen   **ii** the counter?

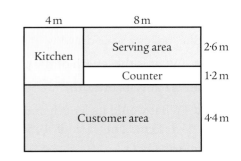

11 Barrey and Wimpett, the housebuilders, have bought a plot of land.
Measurements have been made and entered in a sketch.

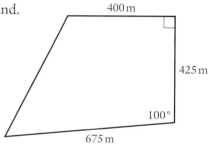

    a Letting 1 cm represent 50 m, make a plan of the land.

    b Find the length in metres of the fourth side of the plot.

    c What is the size of the acute angle in the plot?

Exercise 10.1B

1 An actual length of 5 m is represented on a plan by 25 cm.
What is the scale of the plan? Write your answer as a representative fraction, i.e. in the form '$1:x$'.

2 Work out, as a representative fraction, the scale of a plan where:

    a 10 cm represents 1 m   **b** 4·5 cm represents 9 m

    c 2·8 m is represented by 8 cm   **d** 8·25 m is represented by 16·5 cm.

3 A builder needs to make a plan for a two-room house extension.

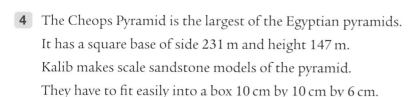

He made a rough drawing containing the accurate measurements.

A plan is to be drawn on A4-sized paper.

The larger the plan, the more accurate the measurements
we take from it will be.

We try to make the most use of the paper available.

    a Using a scale of 1 : 100, make a plan of the extension.
       (All angles are right angles.)

    b Write down the dimensions of the two rooms on the plan.

4 The Cheops Pyramid is the largest of the Egyptian pyramids.
It has a square base of side 231 m and height 147 m.
Kalib makes scale sandstone models of the pyramid.
They have to fit easily into a box 10 cm by 10 cm by 6 cm.

    a Which is a suitable scale for the model, 1 : 250 or 1 : 2500?

    b Write down the length of the base and the height
       of the model.

5 A Boeing 737-800 has a length of 39·47 metres and a wing-span of 34·31 metres.

To make a scale model of the plane, a scale
of 1 cm : 5 m is used.

    a What is the reduction factor?

    b Calculate:
       i the length of the model   **ii** its wing-span.
       (Give both answers to the nearest mm.)

6 The diagram opposite shows an old pattern for a ladies'
skirt ... measurements are given in inches!

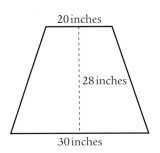

20 inches

28 inches

30 inches

a Make a scale drawing of the skirt using a scale of 1 cm
to represent 4 inches.

b Mark on your scale drawing, in centimetres:

   i the two widths

   ii the length.

c Calculate the area of material needed to make the skirt
in square inches, assuming the actual skirt is made
from two of these pieces.

d A larger size of the skirt is to be made.

   Its measurements are all 10% greater than those for the skirt above.

   i Write down the measurements of this skirt in inches.

   ii Make a scale drawing of the larger skirt, marking on it the two
   widths and the length in centimetres.

e What area of material is now needed? Is it 10% more?

7 Sometimes there is a need to scale up objects.

A snowflake is based on the regular hexagon and is 1 mm across.

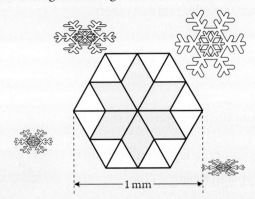

←———— 1 mm ————→

a Letting 1 cm represent 0·25 mm, make a scale drawing of the snowflake.

   The vertices of the inner star are drawn by joining midpoints.

b What is the width of the snowflake on the scale drawing?

8 A whitefly is a tiny insect that feeds on plants.

On average its length is 1·5 mm and its breadth is 0·6 mm.

A scale drawing of the whitefly is made using a scale of
1 cm : 0·2 mm.

What are the dimensions of the whitefly on the scale drawing?

(Note: you are not asked to make a scale drawing of the whitefly.)

9 Yuri is a toymaker who makes Russian dolls.

There are seven dolls in a set, all of different sizes.

The largest six are hollow and can be pulled apart at their middle to allow all the smaller dolls to fit inside the larger ones.

The largest doll is 173 mm tall and 80 mm in diameter at the break.

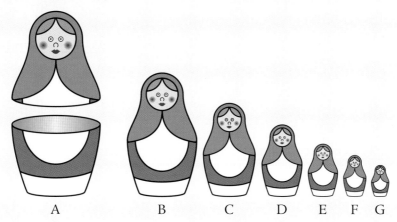

A   B   C   D   E   F   G

Each doll is a $\frac{3}{4}$ reduction of the one before it.

Copy and complete the table, giving your answers to the nearest millimetre.

Doll	A	B	C	D	E	F	G
Height (mm)	173						
Width (mm)	80						

10 The diagram gives the layout of a cottage and its measurements.

   a Using a scale of 1 : 80, draw a plan of the cottage on an A4 sheet of paper.

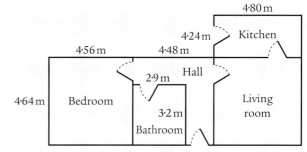

   b On your plan carefully mark, in centimetres, the scaled length and breadth of each room.

11 Carry out this task:

   a Make a rough sketch showing the boundary walls or fences of your school.

   b With a trundle wheel or large measuring tape (50 m or 20 m) carefully measure the lengths of these boundary walls and fences and enter them on your sketch.

   c Use a compass to decide the direction in which each wall runs.

   d Decide on a suitable scale to make a scale drawing of the school grounds.

   e Measure the dimensions of all the school buildings in the school grounds and fit them into your scale drawing in the appropriate places.

10.2 Navigation

Navigators often made scale drawings of journeys.

Each of these scale drawings would be accompanied by either a compass rose to indicate direction, or a north line to give a reference direction for three-figure bearings.

Obviously the scale itself must be given.

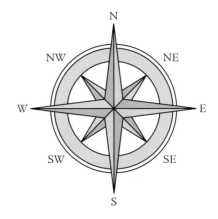

Example 1

From school, the minibus travelled 40 km west, then 24 km north-west to a sports venue.

a Make a drawing of the journey, using a scale of 1 cm to 10 km.

b Work out from the drawing how far the venue is from the school.

a First draw a sketch to organise the data you know:

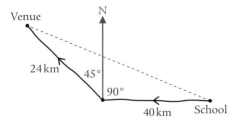

- Plot a point, S, to represent the school.
- Draw a line 4 cm west to represent 40 km.
- Measure an angle of 135° as shown.
- Draw the line 2·4 cm long to represent 24 km, and find V, where the venue is.

b Measure VS ... 5·9 cm.
Scale it up to find the distance of the venue from the school:
actual distance = 5·9 × 10 = 59 km.

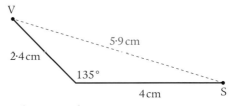

Scale: 1 cm : 10 km

Example 2

The swimming baths (B) are 280 m from the secondary school (S) on a bearing of 142°.

The primary school (P) is 490 m from the baths (B) on a bearing of 250°.

a Make a scale drawing, using a scale of 1 cm : 50 m.

b What is the distance between the primary and secondary schools?

c What is the bearing of the secondary school from the primary school?

a Make a rough sketch to organise the data and your thoughts:

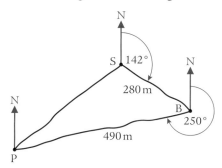

- Plot the point S and a north line from it.
- Using the scale provided, draw SB, and a north line at B.
- Join P and S.
- Use the scale to find actual sizes.

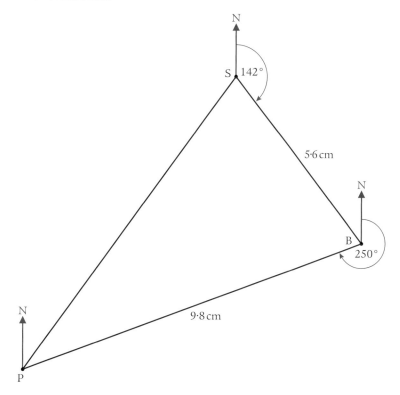

b PS measures 9·7 cm.

⇒ The distance between the schools is (9·7 × 50) m = 485 m.

c At P the angle from the north line clockwise to PS measures 37°.

⇒ The bearing from the primary to the secondary school is 037°.

Exercise 10.2A

1 The radar station at Glasgow Airport (G) shows the positions of planes on a screen.
The airport itself is at the centre of the diagram.

The rings represent distances of 50 km, 100 km and 150 km from the airport.

Plane A is 100 km from the airport on a bearing of 040°.

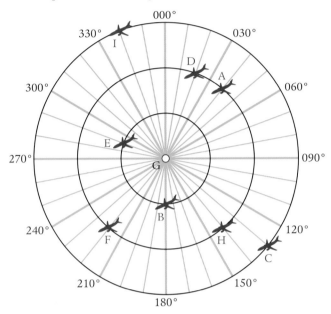

Copy and complete the table:

Plane	A	B	C	D	E	F	H	I
Distance from G (km)	100							
Bearing from G	040°							

2 For each of these journeys:
  i make an accurate scale drawing
  ii from it work out the direct distance back to the starting point.

(Remember to draw a sketch first.)

  a 30 km east then 22 km south. Scale: 1 cm to 5 km.

  b 420 km SW then 350 km west. Scale: 1 cm to 50 km.

  c 24 km NE then 38 km SE. Scale: 1 cm to 4 km.

3 Tom is a delivery driver for the Post Office.

He leaves the depot and drives 250 m in a NW direction to deliver his first parcel.

Then he travels 380 m in a westerly direction to deliver his second parcel.

His next stop is at a house 320 m SE.

He then returns directly to the depot.

  a Using a scale of 1 cm to represent 50 metres, make a scale drawing of Tom's journeys.

  b What is the direct distance of the third delivery point from the depot?

4 Express each of these directions as a three-figure bearing:

a NE   b E   c SE   d S

e SW   f W   g NW   h N.

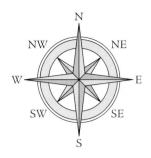

5 Each of these scaled diagrams shows the relative position of two towns.

For each diagram:

   i measure the bearing of town (U) from town (V)

  ii measure the distance on the diagram between the towns

 iii calculate the actual distance between the towns.

a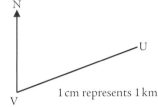

1 cm represents 1 km

b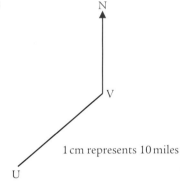

1 cm represents 10 miles

c Scale: 1 : 1 000 000

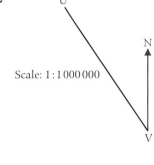

6 Tom and Joss leave school and walk home.

Tom walks 1200 metres on a bearing of 065°.

Joss walks 900 metres on a bearing of 285°.

Using a scale of 1 cm to represent 200 metres, make a scale drawing and from it work out:

a how far apart Tom and Joss live

b what bearing Joss would have to take to get to Tom's house.

7 This scaled diagram shows the course for a cross-country relay running race with three 'legs'.

The runners go from A to B, from B to C and from C to A.

Copy and complete the table:

	Bearing	Distance (m)
Leg 1		
Leg 2		
Leg 3		

Scale: 1 cm represents 500 m

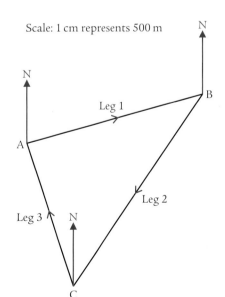

Exercise 10.2B

1 Rhona and Malik are on a week's walking holiday in Perthshire.

They leave their hostel in Aberfeldy and set out for Pitlochry, 18 km away on a bearing of 042°.

On the second day they walk from Pitlochry to Blairgowrie, a distance of 28 km on a bearing of 120°.

The following day they plan to walk back to Aberfeldy.

   a Make a rough sketch of their planned journeys.

   b Using a scale of 1 cm to represent 4 km, make a scale drawing of the walking trip.

   c Use your scale drawing to work out:
     i the distance from Blairgowrie to Aberfeldy
     ii the bearing from Blairgowrie to Aberfeldy.

2 The first four 'legs' of the Beecraigs orienteering course are:

     Leg 1: 300 metres on a bearing of 117°

     Leg 2: 230 metres on a bearing of 150°

     Leg 3: 190 metres on a bearing of 070°

     Leg 4: 200 metres on a bearing of 350°.

   a Make a rough sketch of the course.

   b Use a scale of 1 : 10 000 to make a scale drawing of the four legs.

   c Amy returns directly to the start of the course.
     i Find the bearing for her to return to the starting point of the course.
     ii How far is Amy away from the start of the course at the end of leg 4?

3 There are three airports on the Western Isles: on Barra, on Benbecula and on Lewis at Stornoway.

The rough sketch shows some of the bearings and distances involved.

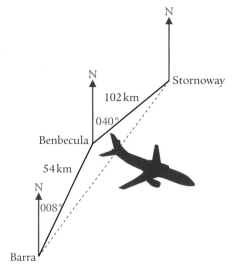

   a Letting 1 cm represent 12 km, make a scale drawing.

   b From your scale drawing find the bearing from Stornoway back to Barra.

   c How much shorter is the journey direct from Barra to Stornoway than the journey via Benbecula?

4 The fishing boat, *Ketch 22*, fishes in the North Sea.

It leaves the harbour at Peterhead (P) and travels 24 km on a bearing of 145° to A.

Three hours later it sails to B on a bearing of 062° for 34 km.

After a successful catch the boat heads back to Peterhead.

a Make a rough sketch showing as much information as you can.

b Using a scale of 1 : 500 000, make a scale drawing of the boat's journeys.

c How far does the boat travel altogether?

d What bearing is required to return from B to Peterhead?

5 A flying doctor is flown in a small plane from island to island to visit her patients.

She does a round trip from Ranald, visiting Coronsey, South Lunaig and North Lunaig before returning to Ranald.

The diagram is drawn to scale.

The actual distance between the two Lunaig islands is 16 km.

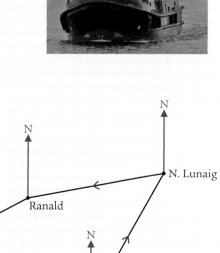

a Determine the scale used in the scale drawing.

b How far does the plane fly altogether? (Give your answer in kilometres.)

c At one point the pilot reports that he is flying on a bearing of 260°.

Which island is he flying towards?

d Measure the bearing the plane takes for each part of the journey.

6 The sketch shows roughly the positions of three farms. Bogend is on a bearing of 205° from Cloverhill.

Lochside is on a bearing of 290° from Cloverhill and on a bearing of 342° from Bogend.

Lochside and Bogend are 19 km apart.

a Make a sketch of the farms, calculating as many angles as you can.

b Make an accurate scale drawing, using a scale of 1 : 200 000.

c Use your drawing to work out whether Cloverhill is closer to Lochside or Bogend, and by how much.

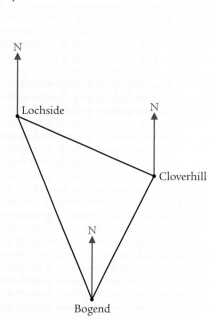

Preparation for assessment

1 On a plan with a scale 1 : 1000, a swimming pool is 2 cm by 1·5 cm.
   What are the actual dimensions of the swimming pool?

2 A football pitch measures 100 m by 70 m.
   What are its dimensions on a plan with a scale of:
   a   1 : 2000
   b   1 : 5000?

3 The first five holes on a golf course are laid out as shown in the scale diagram.
   The scale of the diagram is 1 cm to 100 m.

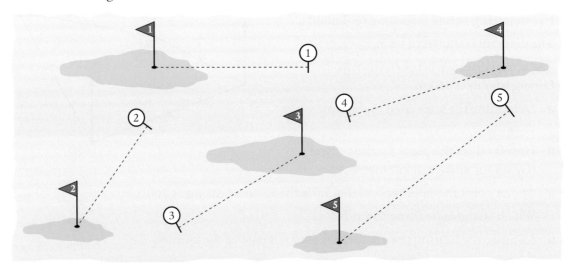

The length of each hole is the distance from the tee to the bottom of the flagstick.
Copy and complete the table:

Hole	1	2	3	4	5
Length of hole, on the plan (cm)					
Length of hole, on the ground (m)					

4 A plane flies the 640 km from Aberdeen to London on a bearing of 170°.
   a   Make a scale drawing, letting 1 cm represent 100 km.
   b   What is the distance between the two places on the scale diagram?
   c   Measure the bearing required for the return journey to Aberdeen.

5 A housebuilder purchases a plot of land.

A rough sketch of the land is drawn and some careful measurements are made and added to the sketch.

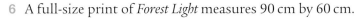

a Using a scale of 1 cm : 40 m, make a scale drawing of the plot of land.

b Use your scale drawing to find the actual length of the third side of the plot.

c Measure the other two angles and write their size on your scale drawing.

6 A full-size print of *Forest Light* measures 90 cm by 60 cm.

The Picture Gallery has decided to make two smaller prints of the painting.

The reduction factor for each is $\frac{2}{3}$.

Find the dimensions of the two smaller prints.

7 The diagram is a scale drawing showing the positions of Bathgate, Broxburn and Linlithgow. The scale is 1 cm to 2·5 km.

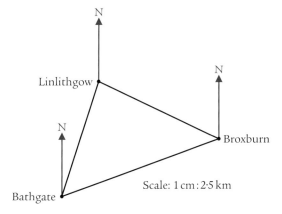

a Use the scale drawing to find the actual distance, as the crow flies, between:
   i Bathgate and Broxburn
   ii Bathgate and Linlithgow
   iii Broxburn and Linlithgow.

b Find the three-figure bearing:
   i from Bathgate to Linlithgow
   ii from Bathgate to Broxburn
   iii from Linlithgow to Broxburn.

8 The map shows the villages around a landfill site (waste dump). The scale of the map is 1 cm represents 5 km.

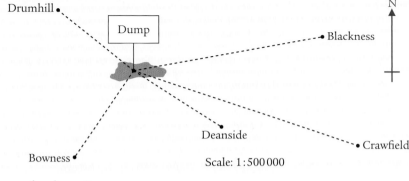

Scale: 1:500 000

a Write the scale in the form 1 : x.

b i Which village is closest to the dump?
   ii What is the actual distance from this village to the dump?

c Write down the actual distances of the other villages from the dump.

d The prevailing wind (the usual wind direction) is from the north-west. Which village is worst affected when it is windy?

e Which village is on a bearing of 215° from the dump?

f Write down the three-figure bearings of the other villages from the dump.

9 Two fishing boats are hunting a shoal of fish.
The *Tom Thumb* is 240 metres due west of the *Mary Mercy*.
On their radar, the *Tom Thumb* picks up the shoal on a bearing of 052°.
The *Mary Mercy* detects the shoal on a bearing of 318°.

a Draw a rough sketch showing the two fishing boats and the shoal.

b Letting 1 cm represent 40 m, make a scale drawing.

c Which boat is closer to the shoal and by how much?

 10 At the beginning of the chapter, you read that you would be asked to design a kitchen.
The project can be tackled by a student working on his or her own, to:

a re-design the kitchen they have at home, or   b design their 'ideal' kitchen.

Alternatively, a pair or group of students could work together in designing their 'ideal' kitchen.

Items for consideration include:

1. dimensions of the kitchen – position of door(s) and window(s)
2. making an accurate scale drawing: perhaps a scale of 1 : 20 is a sensible choice to allow your drawing to be a good size on an A4 sheet of paper
3. position and dimensions of sink, surfaces, units (shelves and cupboards)
4. position and dimensions of cooker/oven, washing-machine, fridge, freezer, dishwasher, etc.
5. placing items from 3 and 4 into your plan
6. costing of items in 3 and 4 – the need to shop around or go online for best deals, weighing up quality versus price
7. costing for floor tiles/carpet/linoleum, wall tiles
8. labour costs for plumber, electrician, joiner, tiler, etc.
9. last, but not least, finding out if there are health and safety regulations regarding things like how close a power socket can be to a sink, or if there are requirements for adequate ventilation of appliances, e.g. a gas cooker.

At the end of the project, you should have a detailed plan of your kitchen and costings of the materials and labour.

11 Making informed choices

⏸ Before we start...

Joe had a star-chart of the Plough, the seven brightest stars in the constellation Ursa Major.

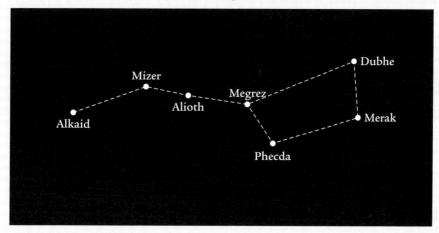

His telescope was programmable.

He could make it move under motor control so that it passed through the field of stars in a straight line.

He wants the line that will make the telescope pass by the stars as closely as possible.

What would be the best path?

A second chart with a grid has been added to help.

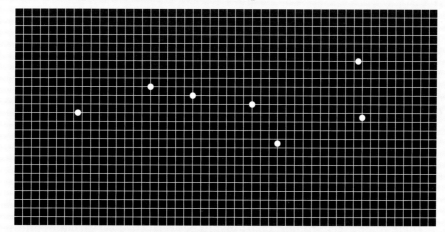

What you need to know

1 a Plot the following set of points:

(2, 1), (1, 2), (3, 3), (4, 5), (5, 4), (6, 6), (7, 8) and (8, 6).

b Describe the pattern this cluster of points makes.

2 a Plot the points in the table.

x	1	2	3	3	5	5	6	7	7
y	2	3	4	5	5	6	7	1	8

b Which two points seem **slightly** out of line?

c Which point is **considerably** out of line?

3 A match manufacturer claims to be 80% confident that the boxes it sells will have at least 48 matches.

How many boxes, in a batch of 200, would the manufacturer expect to have less than 48 matches?

4 Using the terms **highly likely**, **likely**, **unlikely** and **highly unlikely**, describe the following:

a 65% confident

b 15% confident

c 80% confident

d 30%.

11.1 Scatter graphs

In industry, commerce and everyday workplaces, there are many examples of an apparent relationship between two variables.

A good way of investigating how two variables change relative to each other is a scatter graph.

Be careful though, there might seem to be a relationship between the two variables but it does not automatically follow that a change in the first variable will **cause** a change in the second.

If we plotted the sales of ice creams and the number of people with a good tan, we would see an apparent relation. They both would increase together ... but it doesn't suggest that eating ice cream would give you a tan. It does say something about summer habits.

This will be investigated in a later exercise.

Example 1

A construction company compared the average number of bricks laid per hour and the number of years the bricklayer had been fully qualified.

Years qualified	8	2	5	4	5	8	3	6	1
Bricks laid per hour	96	81	92	86	84	99	83	94	79

Show this information in a scatter graph.

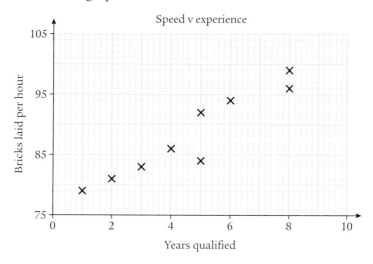

Example 2

The local college held a course for trainee hairdressers on a day-release basis.

Over an eight-week term, the lecturer noted the amount of shampoo each student used and the number of months training they had done.

Training (months)	3	7	9	11	6	7	9	8	6
Shampoo used (ml)	940	730	440	370	680	600	560	580	800

Show this information on a scatter graph.

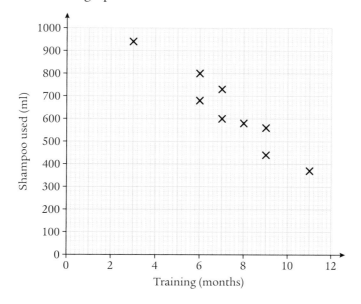

Exercise 11.1A

1 Using suitable scales for the axes, plot the following information on scatter graphs.

a 'Shoe size' (x) and 'Age when first contracted a cold' (y years).

x	2	3	4	5	6	8	9	10	10
y	5	7	4	9	6	3	5	4	7

b 'Days since 1st October' (p) and 'Number of ice lollies sold in shop' (q hundreds)'.

p	22	24	26	28	30	32	34	36	38
q	7	8	6	4	6	5	4	5	3

2 There have been complaints about disruption on the routes between the station and a football ground. Two hours before a big game, a check is made on the number of supporters passing a nearby checkpoint. Nine people from the neighbourhood watching at different times noted the time and the number of supporters passing.

There were s people passing the checkpoint on the dth minute after the start of the count.

d	60	100	70	40	80	50	30	90	110
s	100	200	170	80	150	130	60	60	210

a Draw a scatter graph.

b Which point in the scatter graph looks out of place?

(Note: we refer to such a point as a **rogue value** or **outlier** and would normally disregard it when making decisions.)

3 A shopkeeper noted the average weekly temperature and the total sales for nine weeks.

Week	1	2	3	4	5	6	7	8	9
Temperature (°C)	18	13	19	21	17	18	20	17	15
Weekly sales (£)	530	420	560	600	540	520	510	510	470

a Show this information on a scatter graph.

b Which of the following is most likely to be the item the shopkeeper sold in the shop?
  i Over-coats   **ii** Ice creams   **iii** Pens and pencils

c Give a reason for your answer to **b**.

4 A professional basketball player noted the number of hours he practiced each week for nine weeks. He also recorded the number of points he scored in the league game at the end of each week.

Week	1	2	3	4	5	6	7	8	9
Practice (hours)	20	24	19	23	26	17	21	25	
Points scored	20	22	21	23	25	19	22		20

a Plot the first seven points on a scatter graph.

b Use the first seven points to estimate where the last two points should be. (One part of each point has been given to you.)

5   A building contractor wished to buy pressure hoses to clean up the outside of buildings.
    He tested a variety of hoses comparing the maximum pressure with percentage of
    building cleaned on a first hosing.

Hose type	A	B	C	D	E	F	G	H	I
Maximum hose pressure (psi)	80	130	90	170	120	180	110	160	140
Amount of building cleaned on first hosing (%)	35	64	40	75	50	82	55	75	68

   a   Show this information on a scatter graph.

   b   What else may vary with the hose pressure?

   c   What else will he need to consider when using increased pressure?

6   An artist was advised by a friend that commercially painting pictures
    that cover a large area was a poor option.
    Over a number of months she noted the number of hours spent on
    paintings and the area of canvas used.
    She noted the selling price of each painting and worked out how much
    she made per hour.

Area of picture (m²)	0·3	0·1	0·4	0·2	0·5	0·4	0·6	0·2	0·3
Income from picture (£/h)	13	19	15	15	12	12	8	17	15

   a   Show this information on a scatter graph

   b   What other factors could influence the hourly rate?

Extrapolation and interpolation

Using a scatter graph to make estimations outside the range of the data set is known as **extrapolation**.
Extrapolation should only be used with caution. Answers obtained by extrapolation are not robust and
could be hugely different from actual outcomes.

Estimations made from within the values of the data set are a lot more trustworthy. The process is
called **interpolation**.

Exercise 11.1B

1   A chemist experimented with an additive that increased the strength of wood glues.
    He varied the amount of additive and checked the change in strength of the glue.

Experiment	A	B	C	D	E	F	G	H	I
Amount of additive (ml)	0·7	1·2	1·1	0·9	1·4	1·3	0·8	1·0	1·5
Increase in strength (%)	11	18	18	16	22	22	12	16	24

   a   What percentage increase in strength would you expect if
       1·25 ml of additive was used? (A case of interpolation.)

   b   What percentage increase would you expect if 0·6 ml of
       additive was used? (A case of extrapolation.)

   c   A joint that would normally fail when loaded with 950 kg
       is strengthened by 1·6 ml of additive.
       At what load would you expect this joint to fail? (Extrapolation.)

2 An apprentice chef experimented with cooking a casserole at different temperatures.

He found that the time taken to get the meat succulent also varied.

Experiment	A	B	C	D	E	F
Temperature (°C)	170	150	160	130	140	120
Time (min)	138	150	146	162	155	168

a Show this information on a scatter graph.

b The apprentice was running late. He decided to use a temperature of 210 °C to cook the casserole.

   i How long did he expect the casserole to take to cook?

   ii What condition do you think the meat would be in at the end of this time?

   iii What went wrong with the logic of the apprentice?

11.2 Line of best fit

When we have drawn a scatter graph, it is useful to draw a line through the 'cloud' of points to represent any relationship between the two variables.

There are many lines that might be drawn, but the one that does the job best is called **the line of best fit**. There is a mathematical way of finding this line for any particular data set, but for this course making a judgement 'by eye' is good enough.

The line we choose should run through the cloud of points:

- with roughly the same number of data points above the line as below it
- with a gradient that is similar to the general trend of the 'cloud'
- with the leftmost and rightmost points vertically in line with the ends of the line segment drawn (so that the whole data set is represented).

A line of best fit has been drawn on the graph below.

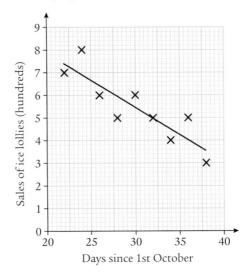

Example 1

A scatter graph has been plotted from a table of information.

Three people draw in what they think is the line of best fit.

Which of the three is the best? Give reasons.

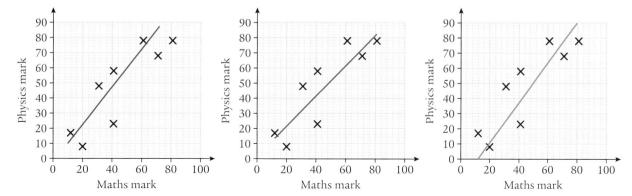

- The red line has the same number of points above as below, but those below are all further away from the line than those above. Also the rightmost data point is not included.

- The blue line has the same number of points above as below. All data points are included. The sum of the distances above the line looks the same as the sum below the line.

- The green line has four points above and four points below. The sum of the distances above the line looks bigger than the sum below the line. By eye, the gradient of the line looks steeper than the cloud of points suggests.

Conclusion: The blue line is the line of best fit.

The line of best fit can be used to estimate values that were not part of the original set of data. (We might estimate that a student who scored 60% in maths would get 62% in physics.)

As long as the line segment is used to get the estimate (interpolation), the estimate can be trusted.

If the line segment has to be extended (extrapolation), any estimate has to be treated carefully.

Example 2

A company worked out the average weekly wage for a group of young trainees.

Age (years)	17	19	16	17	20	19	21	20	22
Wage (£)	170	180	140	160	200	185	205	190	220

a   Show the information on a scatter graph.

b   Draw the line of best fit.

c   Use the line to estimate the average weekly wage for an 18-year-old trainee.

d   By considering the wages of a 12-year-old or a 65-year-old trainee, comment on the extrapolation.

a, b

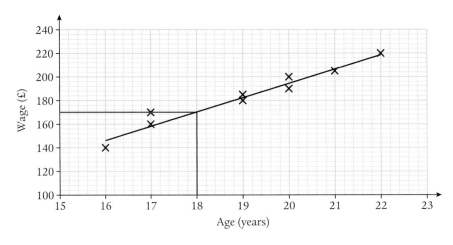

c Using the line of best fit, an age of 18 years corresponds to a wage of £170.

d A 12-year-old is still at school and wouldn't be a trainee.
A 65-year-old might have retired and would possibly not be a trainee.

The relationship between age and wage only holds within the bounds of the chart.

Exercise 11.2A

1 A consumer organisation asked a group of people how much they paid for house insurance.

They also asked them how far their house was from a river.

Household	A	B	C	D	E	F	G	H	I
Distance from river (km)	0·6	0·2	0·4	0·7	0·4	0·6	0·3	0·8	0·7
Cost of insurance (£)	425	550	440	380	500	350	470	275	300

a Show this information on a scatter graph.

b Draw the line of best fit.

c Estimate how much a homeowner who is 0·5 km from the river would expect to pay for house insurance?

2 An apple grower was using a plant food. He wanted to check how good it was.

He gave nine trees a fixed amount of the feed every day. Each tree was given a different daily dose.

He recorded the average weight of the apples harvested from each tree.

Tree	A	B	C	D	E	F	G	H	I
Food per day (ml)	50	80	60	60	90	70	100	80	70
Average weight (g)	143	162	151	148	163	153	161	159	156

a Show this information on a scatter graph.

b Explain why a line of best fit may not be useful.

c What advice would you give to the apple grower?

3 An advertising company considered a possible link between the number of text messages a group of teenagers sent each week and their age.

Teenager	A	B	C	D	E	F	G	H	I
Age (years)	14	16	17	16	18	15	17	13	18
Texts sent each week	68	67	71	49	64	56	54	53	47

a Draw a scatter graph of the data.

b What can you say about the line of best fit here?

c What advice would you give to the advertising agency about targeting age groups?

4 A group of motorists were comparing their annual car insurance premiums.

They thought there was a connection between their premium and their age.

The table shows the ages and annual premiums for the group.

Driver	A	B	C	D	E	F	G	H	I
Age (years)	23	18	24	19	23	20	27	22	19
Annual premium (£)	1000	1250	925	1200	950	1100	480	1050	1250

a Show this information on a scatter graph.

b Which point could be described as a rogue value?

c Ignoring this rogue point, draw a line of best fit.

d Use your line to estimate the insurance premium for a 21-year-old driver.

e Investigate what sort of factors might make the premium for the 27-year-old so out of line.

5 The Inland Revenue noted the earnings and tax paid by a group of self-employed plumbers, for the recent tax year.

They suspected that two of the group were not paying enough tax.

The table shows the results of the group.

Plumber	A	B	C	D	E	F	G	H	I
Earnings (£000)	48	53	36	62	44	57	46	55	42
Tax paid (£000)	6.2	6.4	4.3	7.7	5.8	7.1	4.9	5.8	5.0

a Draw a scatter graph with a line of best fit.

b Which two plumbers might not be paying enough tax?

c The tax man insists that both plumbers pay extra tax so their graph point comes onto the line of best fit.
   i How much extra should each pay?
   ii A 10% surcharge was added to this extra payment for time spent recouping the tax.
       What was the total 'claw back' from both plumbers?

Exercise 11.2B

1 A scientist ran an experiment. He varied the temperature and noted the time it took for a particular chemical reaction to occur.

The graph shows the line of best fit of the data he collected.

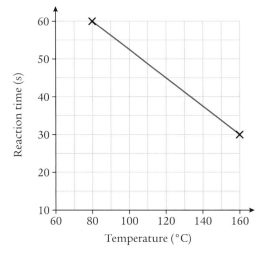

a The actual reaction time at 120 °C was 4% lower than indicated on the line of best fit.

What was the actual reaction time at 120 °C? (Give your answer to nearest second.)

b He extended his line of best fit to extrapolate what may happen at 180 °C.

When he did the experiment, he was surprised to find the actual time was three-quarters of the extrapolated value.

What was the actual reaction time at 180 °C? (Give your answer to the nearest second.)

c A catalyst is a chemical agent that can speed up reactions.

The scientist introduced a catalyst that would reduce reaction times by 0·35 of previous times.

Estimate the reaction time at 160 °C:
  i before the catalyst was added.
  ii after the catalyst was added.

2 A botanist recorded the growth rate of a certain plant when it was kept at different temperatures.

He plotted the data on a scatter graph and drew in the line of best fit. This line indicated that when the temperature was 10 °C, the growth rate was 3 mm per week and when the temperature was 40 °C, the growth rate was 7 mm per week.

A colleague did the same experiment but his line of best fit indicated that at a temperature of 10 °C, the growth rate was 2 mm per week and at 40 °C, the growth rate was 8 mm per week.

At what temperature did both lines of best fit give the same growth rate?

11.3 Correlation

A scatter graph will show any relationship that exists between variables. It also allows us to see if the relationship is strong or weak.

The statistical word we use to describe any such relationship is **correlation**.

The graph will **not** tell us if the relationship is **causal**, i.e. it will **not** tell us if a change in one variable will **cause** a change in the other.

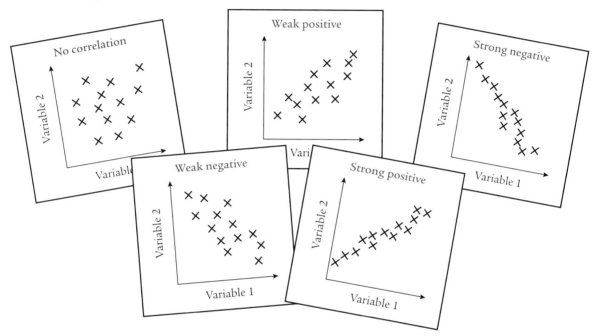

The closer the cluster of points is to forming a straight line, the **stronger** the correlation.

If the points are loosely scattered, we say the correlation is **weak**.

If the line of best fit has a **positive** gradient, we call the correlation positive.

If the line of best fit has a **negative** gradient, we call the correlation negative.

When there is no clear pattern, there is no appropriate line of best fit and no correlation.

Example 1

For each of the following, describe the correlation you think links the two variables.

a   Shoe size and height of a child.

b   Number of workers and time to do a job.

c   The distance you live from the train station and the number of apples you eat.

a   Strong positive correlation. In general, as a child grows you would expect shoe size and height to increase.

b   Strong negative correlation, if the job is getting the hay in from the fields.

Many hands make light work. However, common sense must prevail as the job might be cleaning the inside of a car.

c   No correlation. There is no link between eating apples and distance to the train station.

Example 2

Anna is experimenting with a clockwork car.

Anna knows there is a strong negative correlation between S, the number of turns she gives the key and T, the time in seconds for the toy to go a fixed distance.

The table gives a number of results, but two values are missing, a and b.

What would you estimate to be possible values for a and b?

Trial	A	B	C	D	E	F	G
Number of turns of key, S	5	9	4	10	6	12	b
Time to cover distance, T (s)	9	6	10	5	a	4	6·8

The five known points have been used to create a line of best fit.

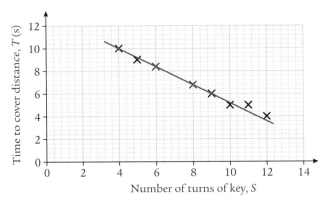

Reading from the chart, 6 turns of the key corresponds to a time of 8·4 seconds ($a = 8\cdot4$).

Also, corresponding to a time of 6·8 seconds is 8 turns of the key ($b = 8$).

Exercise 11.3A

1 When comparing the following, describe the **correlation** you would expect.

Use the terms **strong**, **weak**, **positive**, **negative** and **no correlation**.

  a Average monthly temperature and average monthly heating bill for homes.

  b A person's height and the number of times they go to the cinema.

  c The age of a car and the annual repair bills.

  d The value of a house and the distance it is from the city centre.

  e Average monthly income and amount spent on magazines each month.

2 Two variables form a strong positive correlation.

The line of best fit illustrating the correlation passes through (1, 3) and (6, 8).

Which one of the following points is least likely to be part of the scatter graph on which the line is based?

    (2, 5), (4, 5), (7, 5), (4, 7)

3 A quantity surveyor was asked to price a job to build a long, high brick wall.

He checked the number of bricks laid per hour against how high above ground the bricklayers were working.

Height above ground (m)	0·2	0·5	0·7	0·9	1·2	1·5	1·8	2·0	2·3
Bricks laid per hour	69	66	62	57	53	48	44	41	38

a Show this information on a scatter graph and include a line of best fit.

b How would you describe the correlation?

c Comment on what the line suggests the work rate will be when working 5 m up.

Exercise 11.3B

1 Some of the following pairs of variables are connected causally, i.e. a change in one will cause a change in the other. Identify which are definitely causal.

a The distance to the horizon and the height you climb a tower.

b The temperature and the number of people in a room.

c Shoe size and height.

d The number of times a machine breaks down and the number of moving parts in the machine.

e Your age and your score in a general knowledge quiz.

f The time you spend on your homework and the score you get in the exam.

g The amount of Perrier water you drink and your score in the French exam.

h The number of electrical appliances in your house and your electricity bill.

2 a Plot the points given in the table.

x	2	3	5	6	7	8	9	11
y	8	4	8	3	8	2	6	2

b There is a very weak relation between x and y. Describe it.

c Draw a straight line through (2, 7) and (11, 3). This is the line of best fit.

d The y-value for any point below the line is to be increased by 50%. The y-value for any point above the line is to be decreased by 25%.

    i Plot the new points.   **ii** How would you describe the correlation now?

11.4 Chance and uncertainty in the workplace

We already know that the probability of:

- getting a head, when a fair coin is tossed, is $\frac{1}{2}$

- drawing a spade from a normal pack of cards is $\frac{1}{4}$

- picking a domino at random from a double six set, that has more than eight spots in total, is $\frac{6}{28}$ or $\frac{3}{14}$.

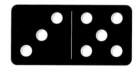

This section of work will look at making informed choices, based on the probability of an event. When the mathematical probability may be difficult to calculate, a verbal description will be used.

Example 1

Lillian is hoping to open a ladies hair styling salon. She wants to know how profitable it might be.

What questions should she research before making her decision ... and why?

The major questions would be:

- How many hairdressers are already in the area? (This will affect the probability of someone choosing her salon.)

- How often do the females in the area go to the hairdresser? (This will affect the probability of someone wanting the hairdresser.)

- What, on average, are they willing to spend on a visit? (This will affect the rate of earnings.)

- What are the other hairdressers charging? (Unless she is competitive, the probability of customers visiting her drops.)

- How much will be needed for rental, business rates and other charges? (This will affect her profits.)

- How many hairdressing businesses have closed in the area over the last few years? (This will be an indication of the probability of success.)

- Is there a suitable location available to use as a premises? (A poor location will drastically reduce the probability of getting customers.)

Example 2

Brian works for a company where time-keeping is important.

Employees who are late for work are given up to three warnings.

Being late four times, in any calendar year, results in dismissal.

By 1st August, Brian has been late twice.

The bus he catches to work usually runs late 3% of the time.

a What are the calculations Brian has to make?

b What possible lifestyle changes does he need to consider?

a Working days remaining this year = 100 (holidays have been taken out).

Bus runs late 3% of the time. 3% of 100 = 3 ... the bus will be late three times by the end of the year.

If lateness of the bus is spread evenly, he will be late twice in the next 66 working days.

He is **very likely** to be dismissed by early November, if not even earlier.

b Brian needs to have a serious look at his time management and think of other travel arrangements ... an earlier bus possibly ... and soon. Leaving the decision until later is very dangerous.

The bus is late three in every 100 trips ... but this could happen in the next three journeys.

Exercise 11.4A

1 The quality assurance process in a factory that makes light bulbs is such that they are 98% confident that any bulb leaving the factory will work effectively.

The bulbs are sold in packs of two.

A hardware store ordered 500 packs.

   a How many defective bulbs would be expected in the order?

   b How likely is it that someone buying a pack of two would find neither worked?
Explain your answer using terms such as highly likely, not very likely, etc.

2 Tom wants to start up as a self-employed window cleaner.

His research indicates that he could expect 3% of a town of 6000 to become customers.

He charges £4·50 and cleans the windows every second week.

   a What would he expect to earn each week?

   b If he raises his charge to £5 he knows this will result in losing 1 in 10 of his customers. What advice would you give Tom about the price rise?

3 The local fishmonger sells all his fish every day.

He orders an extra 10 kg a day.

He buys the fish at £4 a kg and sells it at £7 a kg. His shop is open five days a week.

On four of the days he sold all the extra stock.
On one day he sold none of it and all the extra stock had to be thrown out.

How much extra income per week should the extra 10 kg a day bring in?

4 A company that makes kit houses orders wood from the local timber yard.

The company is informed there is a 90% chance that soon timber prices will rise by 10%.

Timber is ordered in bulk every six months and this order usually amounts to £80 000.

If the company orders now, the timber will be at the old price. To do this it needs a bank loan with an interest repayment of £2500.

What advice would you give to the chief buyer of the kit house company?
(Show all calculations to support the advice.)

5   The Kwik Kup coffee shop makes coffee from ground beans.

The table shows the trend in the cost of beans to the shop and the price they charged for a cup of coffee.

Year	2008	2009	2010	2011	2012
Cost of beans (£ per tonne)	2000	2100	2150	2250	2300
Cost of a cup of coffee (£)	1·60	1·75	1·80	1·90	2·00

Over this period, the amount of cups of coffee sold has remained fairly constant.

a   What do these figures indicate as far as profits are concerned?

b   What assumptions have you made?

Exercise 11.4B

1   The local kebab shop has a regular client base of 400 customers per week.

They spend an average of £12 per week.

a   What is the total weekly takings for the kebab shop?

b   The owner is considering raising prices by 10%.

He has been advised that every 5% his prices increase will result in a loss of 10% of customers.

How would this affect the average weekly spend of a customer?

c   How would this affect the total weekly takings?

2   Mr Di Angelo kept records on how his ice cream sales had varied with temperature.

The table shows the data he had.

Maximum daily temperature (°C)	16	17	18	19	20	21	22	23	24
Average daily sales (litres)	35	40	42	53	55	58	65	65	68

He presently has 45 litres of ice cream in stock and his weekly delivery is due tomorrow.

The Met Office has just issued the following temperature forecast for the next seven days.

Day	1	2	3	4	5	6	7
Maximum temperature (°C)	21	23	19	21	23	24	22
Degree of confidence (%)	80	50	90	60	80	80	60

The first column indicates there is an 80% degree of certainty that the temperature will reach 21 °C.

How many litres of ice cream would you advise Mr Di Angelo to order for the coming week? (Show all your working and give reasons.)

11.5 Personal chance and uncertainty

This is an extension of the last section, with personal events being considered.

The concepts and skills are the same but the decisions are based on different contexts.

Example 1

The Smart family have been advised that their old central heating boiler is costly to run.

It is 80% certain that the boiler will need major repair work, or replacing, in the next two years.

The annual heating bill using the old boiler is £650.

A new boiler will cut the heating costs by 30% a year.

The cost of supplying and fitting the new boiler will be £1500. This £1500 can be split over 10 years and added to the heating bill. This will incur an annual interest of £70.

What options do the Smart family need to consider?

Consider the annual cost of the new boiler.

- 30% of £650 means a saving on heating bills of £195 per year.

- Annual payment for new boiler $= \dfrac{£1500}{10} + £70 = £220$.

- Actual cost after savings on fuel bill $= £220 - £195 = £25$ per year.

There is an 80% chance that there will be a repair bill within the next two years (80% = **very likely**), so replacing the boiler is almost self-financing.

Exercise 11.5A

1 Alice was considering the cost of house and contents insurance.
Her annual insurance premium is £285.
Statistics show there is a 1 in 20 chance of making a claim.
The average claim made is £3800.

    a **i** If these statistics prove to be correct for Alice, how much extra will she expect to pay to the insurance company, over a twenty-year period, after she has made one average claim?

        ii How much is this per year?

    b Discuss/investigate the various factors she might take into consideration.

2 The Scott family was looking to buy a new TV.
The electrical retailer strongly suggested they take out an extended five-year warranty.
The cost of the new TV was £235. The cost of buying the extended warranty was £165. The actual warranty is for one year only.
The TV manufacturer was 95% confident that all their products would function effectively for at least seven years.
Explain why you would or would not take out the extended warranty. (Your decision should be based on the data given.)

3 Fifty per cent of persistent smokers will die from a smoking related disease. Of these deaths, 1 in 4 will be caused through lung cancer.

    a Every year 150 000 young people in the UK start smoking. If all of them persist in smoking, how many will be expected to die from lung cancer?

    b Use a search engine on the internet to find out what chronic obstructive pulmonary disease (COPD) is and how smoking is influencing trends.

4 In a recent TV programme, a health expert made the following statement:
'If alcohol was just starting to be used by the public, almost all world governments would ban it immediately'.
He based these comments on the statistics for alcohol related diseases and crimes where alcohol played a contributory part.

    a Research the trends in alcohol consumption over the last few years. (You could make use of the internet to find suitable data on the government's own statistical site.)

    b Find out how much alcohol related diseases are estimated to cost the NHS, each year.

    c Discuss the case for making alcohol a prescribed substance, the same as class A drugs.

Exercise 11.5B

1 Recent statistics show that 18% of people involved in recorded road accidents, in Scotland, were in the 16 to 22 age group.

Young male drivers in the age-group 17 to 25 years are one-and-a-half times more likely to be involved in a car accident than drivers in any other age-group.

People in their late teens and early twenties are usually at a peak as far as eyesight, hearing, reaction time and general physical fitness are concerned.

    a How can you explain the contradiction in the statements above?

    b Explain the link between the statement on car accidents and car insurance for young drivers.

    c Research sex equality legislation and the impact it has on young female drivers in the age-group 17 to 25 years.

    d By extending this research to sex equality legislation and life insurance, discuss how the statistics would indicate that women are disadvantaged by recent legislation.

2   The average life expectancy for a man and a woman in 1900 was 47 and 50 respectively.

By 2000, the average life expectancy had risen to 77 and 81.

a   By extrapolation, what figures would you expect for 2050.

b   These are average life expectancy figures and are calculated from all deaths.

    Research child mortality rates in 1900 and 2000.

c   Explain why the child mortality figures may skew the extrapolation in **a**.

Preparation for assessment

1   Nine children were asked to count the number of steps it took them to cover 10 m.

Child	A	B	C	D	E	F	G	H	I
Age	5	9	7	11	8	6	10	4	12
Number of steps taken	24	19	23	17	20	22	16	26	15

a   Draw a scatter graph.

b   Describe the correlation between age and number of steps taken.

2   An advertising agency monitors a group of companies that supply a similar product.

They note the number of times the company has an advert on TV and the sales, for a particular month.

Company	A	B	C	D	E	F	G	H	I
Times advert appeared	23	17	19	23	18	20	24	22	19
Sales for month (£000)	10·4	8·5	9·5	11·1	9·2	9·4	11·3	10·4	9·0

a   Show this information on a scatter graph.

b   Draw in a line of best fit.

c   What sales would you expect with 21 adverts in the month?

3   Recent annual statistics showed that 20 people died as a result of accidents at work.

The same set of statistics indicated there were 2 500 000 people in work.

a   Calculate the probability of an employee, chosen at random, being killed at work.

b   Name a job that you consider dangerous and one that you think carries little danger.

c   Comment on your answer to **a** in relation to the safe job you mentioned in **b**.

d   Name three items that you should wear on a construction site, for your safety.

4 Online gambling is growing at a spectacular rate.

The companies involved spend a fortune on advertising.

One of these companies made annual profits of more than £100 million.

a A conservative estimate is that overheads, advertising and profits amounted to £300 million.

Where does this £300 million come from?

b For a certain company, this £300 million represented 12·5% of the total bets taken.

What was the total value of bets taken by this one company?

5 Remember Joe's star-chart? It was a star-chart of the Plough, the seven brightest stars in the constellation Ursa Major.

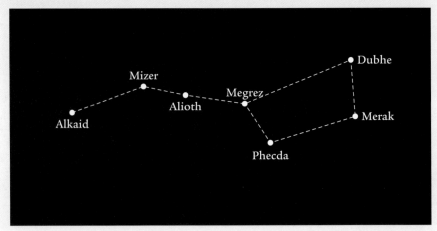

Joe's telescope was programmable.

He could make it move under motor control so that it passed through the field of stars in a straight line.

He wants the line that will make the telescope pass by the stars as closely as possible.

What would be the best path?

A second chart with a grid has been added to help.

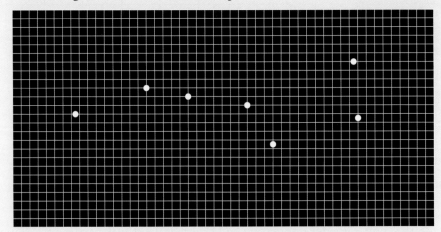

⏸ Before we start...

With three gallons of petrol in his fuel tank, Tony drove his car at a steady 70 mph around a test circuit.

He drove for $2\frac{1}{2}$ hours before the car ran out of petrol.

Tony put another three gallons of petrol in the car and then drove the car round the circuit at a steady speed of 50 mph.

It was $3\frac{3}{4}$ hours later that the car came to a halt.

In which trial did the car travel further?
By how much?

▶ What you need to know

1   Express each of these times as times on the 12-hour clock with a.m. or p.m.

    a   03 40   **b**   14 50   **c**   22 05   **d**   10 53

2   Express these times as times on the 24-hour clock:

    a   3.15 a.m.   **b**   5.58 p.m.   **c**   11.03 p.m.   **d**   12.30 a.m.

3 The Boeing 737 leaves Glasgow Airport at 6.40 a.m. and arrives in Tenerife at 11.05 a.m. How long does the flight take?

4 Samir and Val are planning their journey from Glasgow to Kilwinning.

They need to arrive in Kilwinning just before 9 a.m. and be back in Glasgow before 7 p.m.

Glasgow Central	d	•	•	0600	0615	0630	0645	0700	0715	0730	•	0800	•	0815	0830	0834	0845	0900
Paisley Gilmour St	d	•	•	0612	0626	0641	0656	0711	0726	0741	•	0811	•	0826	0841	0845	0856	0911
Johnstone	d	•	•	0614	0630	0645	0700	0715	0730	0745	•	0815	•	0830	0845	0849	0900	0917
Miliken Park	d	•	•	•	0633	•	0703	•	0734	•	•	•	•	0833	•	0852	0903	•
Howwood	d	•	•	•	0636	•	0706	•	0737	•	•	•	•	0836	•	0855	•	•
Lochwinnoch	d	•	•	•	0640	•		•	0741	•	•	•	•	0840	•	0859	•	•
Glengarnock	d	•	•	•	0645	•	0713	•	0746	•	•	0824	•	0845	0854	0904	0911	•
Dalry	d	•	•	•	0649	•	0717	•	0750	•	•	0828	•	0849	0858	0908	•	•
Kilwinning	d	•	•	0628	0654	0659	0723	•	0754	0800	•	0833	•	0854	0903	0912	0918	0931
Stevenston	d	•	•	•	0657	•	0727	•	0758	•	•	•	•	0857	•	0916	0921	•
Saltcoats	d	•	•	•	0700	•	0729	•	0800	•	•	•	•	0900	•	0918	0924	•
Ardrossan South	d	•	•	•	0702	•	0731	•	0802	•	•	•	•	0902a	•	0920	0926	•
Ardrossan Town	d	•	•	•	•	•	0735	•	•	•	•	•	•	•	•	0924	•	•
Ardrossan Harbour	d	•	•	•	•	•	0737	•	•	•	•	•	•	•	•	0926	•	•
West Kilbride	d	•	•	•	0708	•	•	•	0808	•	•	•	•	•	•	•	0931	•
Fairlie	d	•	•	•	0713	•	•	•	0814	•	•	•	•	•	•	•	0935	•
Largs	a	•	•	•	0720	•	•	•	0820	•	•	•	•	•	•	•	0944	•

Largs	d	•	1650	•	•	1735	•	•	•	•	•	•	•	1853	•	•	•
Fairlie	d	•	1655	•	•	1740	•	•	•	•	•	•	•	1858	•	•	•
West Kilbride	d	•	1700	•	•	1745	•	•	•	•	•	•	•	1903	•	•	•
Ardrossan Harbour	d	•	•	•	•	•	•	1800	•	•	•	•	•	•	1928	•	•
Ardrossan Town	d	•	•	•	•	•	•	1803	•	1832	•	•	•	•	1930	•	•
Ardrossan South	d	•	1706	•	1742	1751	•	1806	•	1834	•	•	1909	•	1934	•	•
Saltcoats	d	•	1708	•	1744	1753	•	1808	•	1836	•	•	1911	•	1936	•	•
Stevenston	d	•	1711	•	1747	1756	•	1811	•	1839	•	•	1914	•	1939	•	•
Kilwinning	d	1705	1715	1736	1750	1800	1804	1816	1837	1843	•	1904	1918	1937	1943	•	2004
Dalry	d	•	1720	1741	1755	1805	1809	•	1841	1847	•	•	•	•	1947	•	•
Glengarnock	d	•	1724	•	1759	•	1813	1823	•	1851	•	•	1925	•	1951	•	•
Lochwinnoch	d	•	1729	•	•	•	1818	•	•	1856	•	•	•	•	1956	•	•
Howwood	d	•	•	•	•	1814	•	1830	•	1900	•	•	•	•	2000	•	•
Miliken Park	d	•	1734	•	1807	•	•	1833	•	1903	•	•	1933	•	2003	•	•
Johnstone	d	1719	1736	1752	1809	1818	1824	1835	1852	1906	•	1918	1935	1950	2006	•	2018
Paisley Gilmour St	d	1724	1741	1758	1814	1823	1829	1840	1857	1911	•	1923	1940	1957	2011	•	2025
Glasgow Central	a	1736	1755	1809	1825	1834	1842	1852	1908	1922	•	1934	1953	2008	2022	•	2036

a Which trains should they catch to give them the most time in Kilwinning?

b How long will they be able to stay in Kilwinning?

12.1 Time intervals – duration

A time interval can be calculated by subtracting the **start time** from the **end time**.

We have to remember that 60 minutes make one hour.

Example 1

A train leaves Waverley Station in Edinburgh at 05 40. It arrives in Kings Cross, London at 10 05.

How long did the journey take?

Subtract the **start time** from the **end time**:

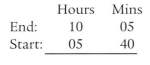
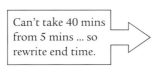

	Hours	Mins
End:	10	05
Start:	05	40

Can't take 40 mins from 5 mins ... so rewrite end time.

	Hours	Mins
End:	09	65
Start:	05	40
	4	25

If you are working with the 12-hour clock and an event crosses noon or midnight, then remember that the end time is actually 12 hours more than is given.

Example 2

A charity Telethon is timetabled to start at 7.25 a.m and end at 11.15 p.m.

What is the duration of the Telethon?

The event crosses noon ... so add 12 hours to the end time:

End time becomes 11.15 + 12.00 = 23.15

Then subtract the **start time** from the **end time**:

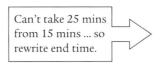

	Hours	Mins
End:	23	15
Start:	07	25

Can't take 25 mins from 15 mins ... so rewrite end time.

	Hours	Mins
End:	22	75
Start:	07	25
	15	50

The Telethon lasts for 15 hours and 50 minutes.

If you are working with the 24-hour clock and an event crosses midnight, then remember that the end-time is actually 24 hours more than is given.

Alternative method

The most awkward time intervals to calculate can be the ones that cross midnight.

If necessary, use these time lines to help you.

Work out the time interval **before midnight**, the time interval **after midnight**, then add them together.

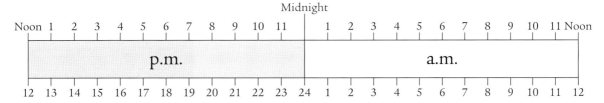

Example 3

The Midnight Swim session at the local swimming baths begins at 10.30 p.m. and ends at 1.45 a.m.
How long does the Midnight Swim session last?

     10.30 p.m. to 12 (midnight) = 1 hour 30 minutes

     12 (midnight) to 1.45 a.m.   = 1 hour 45 minutes

     Total   = 2 hours **75 minutes** = 3 hours 15 minutes

The Midnight Swim session lasts 3 hours and 15 minutes.

Example 4

Pat is a nurse at the local hospital. When she is on night duty, her shift lasts from 20 45 until 08 15.
How many hours and minutes does her shift last?

     20 45 to 24 00 (midnight) = 3 hours and 15 minutes

     00 00 (midnight) to 08 15 = 8 hours and 15 minutes

                   Total = 11 hours and 30 minutes

Pat's shift lasts 11 hours and 30 minutes.

Exercise 12.1A

1 Calculate the duration of each time interval.

    a 9 a.m. to 7 p.m.   **b** 1.30 a.m. to 8.15 p.m.

    c 10.20 p.m. to 4.10 a.m.   **d** 9.37 p.m. to 8.25 a.m.

2 Express the time interval between each pair of times in hours and minutes.

    a 06 50 and 12 10   **b** 14 56 and 23 07

    c 21 40 and 09 10   **d** 15 42 and 00 18

3 The school day at Deanburn High School begins at 8.50 a.m. and ends at 3.35 p.m.

    a How long is the school day?

    b How long is the school day if the morning interval of 15 minutes and lunchtime of 40 minutes are not included?

    c How long is the school week (morning intervals and lunchtimes not included)?

    d Repeat **a**, **b** and **c** for your own school.

4 **a** Sunrise is at 07 35 and sunset is at 20 08.

       How long is it from sunrise to sunset?

    b On the same day, the Moon rises at 21 28 and sets at 06 13.

       For how long is the Moon visible in a cloudless sky?

5 How long does each of these coaches take to travel from Aberdeen to London?

	Aberdeen (depart)	**London (arrive)**
a	04 40	17 30
b	07 05	20 30
c	19 15	07 05

6 Which of the overnight coaches from Glasgow to London is:

a quickest   **b** slowest?

Glasgow (depart)	**London (arrive)**
22 30	06 50
22 45	07 25
23 00	10 20

7 Darren and Molly travel by car from Edinburgh to Glasgow to deliver a parcel to a friend.

They then return to Edinburgh.

The distance–time graph gives details of their journey.

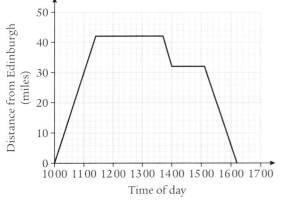

a How long did it take them to get to their friend's house?

b How far did they travel to reach their friend's house?

c How long did they spend there?

d On their return journey they got caught up in a traffic jam.
  i How long were they held up?
  ii How far from their friend's house was the hold-up?

e At what time did they arrive back in Edinburgh?

8 Does your strategy for working with **clock times** also work with **calendar times**?

Stuart's date of birth is 09/04/1999. The date of birth of his friend Sam is 30/07/1999.

How many weeks is Stuart older than Sam?

Exercise 12.1B

1 **a** The Sun rises at 06 48 and sets 13 hours and 25 minutes later.

When does the Sun set?

b The Moon rises at 18 32 and sets 7 hours and 42 minutes later.

When does the Moon set?

2 **a** Lanzarote is in the same time zone as Edinburgh.

Flight RY476 from Edinburgh to Lanzarote arrived in Lanzarote at 23 08.
The flight had taken 4 hours and 10 minutes.

When did the plane leave Edinburgh?

b Madeira and Glasgow are in the same time zone.

Flight TC891 from Madeira to Glasgow landed in Glasgow at 03 27
after a flight that lasted 4 hours 35 minutes.

When did the plane leave Madeira?

3 George and Sal are long-distance lorry drivers who work for the same company. One night George
drives from Paisley to Birmingham as Sal drives from Birmingham to Paisley. The distance–time
graph describes their journeys.

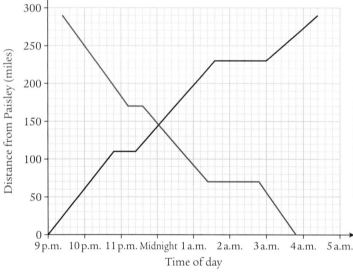

a What is the distance between Paisley and Birmingham?

b Whose journey is illustrated by the red line?

c For how long did George drive before he first stopped?

d For how long did he stop?

e How far from Paisley was Sal when she first stopped?

f **i** When did they pass each other?
ii How far were they from Paisley when they passed each other?

g How long did the whole journey take:
i George **ii** Sal?

4 The Sun rises at different times in different places around the world as the Earth spins.

The first city on the Earth to welcome the new day is Auckland in New Zealand.

As the Earth spins, the new day gradually begins in places to the west of Auckland.

Below is a selection of cities across the world and the time it is there when it is noon in Edinburgh.

City	Time
Anchorage	3.00 a.m.
Los Angeles	4.00 a.m.
New York	7.00 a.m.
Edinburgh	Noon
Oslo	1.00 p.m.
Athens	2.00 p.m.
Mumbai	4.30 p.m.
Beijing	7.00 p.m.
Melbourne	9.00 p.m.
Auckland	11.00 p.m.

a How many hours ahead of Edinburgh is:
  i Auckland   ii Beijing   iii Mumbai?

b How many hours behind Edinburgh is:
  i Los Angeles   ii Anchorage   iii New York?

c When it is 3.35 p.m. in Edinburgh, what is the time in:
  i Melbourne   ii New York   iii Mumbai?

d Use your knowledge of geography to decide if the time in each of these cities is ahead or behind the time in Britain.
  i Tokyo   ii Athens
  iii Chicago   iv Rio de Janeiro

5 The plane from Glasgow to New York leaves Glasgow at 09 50.
The flight time is 6 hours and 40 minutes.

When does the plane arrive in New York (use the table in question 4):

a British time

b New York time?

6 A flight from Glasgow to Mumbai left Glasgow at 21 45.

It landed at Mumbai 10 hours and 50 minutes later.

At what time did it land at Mumbai (use the table in question 4):

a British time

b Mumbai time?

12.2 Rate and speed

A **rate** tells you how one measurement changes with respect to another.

The word 'per' (meaning 'for each') is used to separate the units of the two measurements.

Reading the units that a rate is given in will give you a clue about how to find it. For example,

miles per litre ... divide miles by litres

litres per kilometre ... divide litres by kilometres

kilograms per minute ... divide kilograms by minutes.

By law, vendors have to display their rates.

Notice that on the petrol pump the rate 'Pence per Litre' is visible.

Example 1

A bricklayer lays 240 bricks in 10 minutes.

Calculate the rate at which the bricklayer is working, in bricks per minute.

Think: bricks per minute ... divide number of bricks by number of minutes.

The rate at which the bricklayer is working is:

$$\frac{240}{10} = 24 \text{ bricks per minute.}$$

Example 2

The fastest growing plant is a variety of bamboo.

One grew 852 centimetres in eight days.

a   Calculate its rate of growth in:
   i   centimetres per day
   ii centimetres per hour
   iii hours per centimetre.

b   What is the relation between answer **ii** and answer **iii**?

a **i** Rate is centimetres per day. Think: $\dfrac{\text{cm}}{\text{days}} = \dfrac{852}{8} = 106\cdot5$ cm per day.

    ii Rate is centimetres per hour. Think: $\dfrac{\text{cm}}{\text{hours}} = \dfrac{106\cdot5}{24} = 4\cdot44$ cm per hour (to 2 d.p.).

    iii Rate is hours per centimetre. Think: $\dfrac{\text{hours}}{\text{cm}} = \dfrac{24}{106\cdot5} = 0\cdot23$ hours per centimetre (to 2 d.p.).

b The product of answer **ii** and answer **iii** is 1 (allowing for rounding errors).

Speed

Speed is a special rate.

It is the rate at which **distance** changes with **time**.

When you go on a journey it is unlikely you will travel at a constant speed for the whole journey.

If a train takes one hour to travel the 48 miles from Girvan to Glasgow, its speed will vary from 0 mph, when it stops at a station, to about 70 mph when it is between stations.

We would say the **average speed** of the train is 48 mph.

$$\textbf{Average speed} = \frac{\textbf{total distance covered}}{\textbf{total time taken}}$$

$$S = \frac{D}{T}$$

where S = speed or average speed, D = total distance travelled, T = total time taken.

Example 3

Molly drives the 647 km from Glasgow to London in seven hours.

Calculate her average speed for the journey.

Average speed, $S = \dfrac{D}{T} = \dfrac{647}{7} = 92\cdot4$ km per hour (or km/h), to 3 s.f.

Example 4

Tariq's journey of 145 miles from Dumfries to Dundee took 3 hours and 42 minutes.

Calculate his average speed for the journey in mph.

We want to find the average speed in mph, so we need the distance in miles and the time in hours.

First change 42 minutes into a fraction of an hour (remember, there are 60 minutes in an hour):

$$42 \text{ minutes} = \frac{42}{60} \text{ hour} = 0.7 \text{ hour.}$$

Total time is 3·7 hours.

So average speed $= S = \dfrac{D}{T} = \dfrac{145}{3.7}$ mph $= 39.189189... = 39.19$ mph (to 2 d.p.).

Tariq's average speed was 39·19 mph (to 2 d.p.).

Exercise 12.2A

1 Tommy stacks the shelves in a supermarket.

He works 20 hours each week and his pay is £110.

Calculate his rate of pay per hour.

2 Ann, Brenda and Chris work in an office.

Ann typed a 464-word document in eight minutes.

Brenda typed a 270-word letter in five minutes

Chris typed a 343-word report in seven minutes.

a Calculate the number of words per minute each person typed.

b Who typed at: **i** the fastest rate   **ii** the slowest rate?

3 A swimming pool holds approximately 100 000 litres of water.

To make repairs to the bottom of the pool, it was emptied of all its water. This took one-and-a-half hours.

a Calculate the rate at which the water was drained, in:
   i litres per hour
   ii litres per minute.
   (Give your answers correct to 3 s.f.)

b Calculate the rate the pool emptied in minutes per litre.

4 Fergus put leaflets through the letterboxes of the houses in his neighbourhood.

It took him $2\frac{1}{2}$ hours to deliver 420 leaflets.

Calculate his rate of delivery in leaflets per hour.

5 **a** When you are resting, your heart pumps about 340 litres of blood every hour.

What rate is this in: **i** litres/minute   **ii** litres/day? (Note: the symbol '/' is the same as 'per'.)

b At what rate does your heart pump blood during heavy exercise, given that the rate increases six-fold?

Give your answer in: **i** litres/hour   **ii** litres/minute.

6 Chloe cycled 16 miles in three hours. Debbie cycled 21 miles in four hours.

    a Whose average speed was greater?

    b By how much?

7 **a** The world record for the men's 50-metre freestyle swim is 20·91 seconds.
Calculate the average speed in m/s.

    b The world record for the men's 100-metre freestyle swim is 46·91 seconds.
Calculate the average speed in m/s.

    c Which distance produced the greater average speed? By how much?

8 The winners' times in some of the athletics events at the London 2012 Olympic Games are given in the table.

Women			Men		
Event	Time of winner	Rate (m/s)	Event	Time of winner	Rate (m/s)
100 metres	10·75 s	A	100 metres	9·63 s	E
200 metres	21·88 s	B	200 metres	19·32 s	F
400 metres	49·55 s	C	400 metres	43·94 s	G
800 metres	1 min 56·19 s	D	800 metres	1 min 40·91 s	H

    a Calculate the rate or average speed of each winner in metres per second (the values **A** to **H**).
Give your answers correct to 2 decimal places.

    b Comment on your answers.

9 Peter and Helen took their children on a walk along the path from Galashiels to Melrose.

A signpost gives certain distances in miles.

    a What is the distance between the towns according to the sign?

    b They set out at 10.15 a.m. and arrived in Melrose at 11.45 a.m.
What was their average speed?

Example 5

A travelling salesman went from Dundee to Queensferry (48 miles) at an average speed of 50 mph.

He then went from Queensferry to Dalkeith (19 miles) at a speed of 40 mph.

What was his average speed for the whole journey?

We need the total distance: $48 + 19 = 67$ miles.

We need the total time: leg 1 + leg 2 $= \dfrac{48}{50} + \dfrac{19}{40} = 0.96 + 0.475 = 1.435$ hours.

So average speed $= \dfrac{67}{1.435} = 46.7$ mph (to 1 d.p.).

Exercise 12.2B

1 Amir earns £160·80 every week for working 20 hours in a supermarket.

Billy is paid £193·20 for working 35 hours a week in a laundry.

Charlie works in a hotel and is paid £202·50 each week for working 37·5 hours.

Calculate the hourly rate of pay of each boy.

2 Emma can knit 75 stitches in 5 minutes.

   a Calculate the rate at which Emma can knit, in stitches per minute.

Emma is knitting a jumper with 90 stitches in each row of the back.

   b Calculate the rate at which she can knit the back in rows/hour.

3 Change each of these times to hours:

   a 1 hour 30 minutes   **b** 4 hours 18 minutes   **c** 2 hours 27 minutes.

4 Change each of these times to hours, giving your answers correct to two decimal places:

   a 3 hours 20 minutes   **b** 5 hours 16 minutes   **c** 4 hours 55 minutes.

5 Work out the average speed for each of these.

   a A hill walker covered 22 km in 5 hours.

   b A cyclist managed 108 km in 3 hours.

   c A car travelled 472 km in 4 hours 30 minutes.

   d A jet flew 4850 km in 2 hours 15 minutes.

     (Give the answers to **c** and **d** to the nearest whole number.)

6 A driver was wishing to go from Stirling to Kirkcudbright.

Using Google Maps, he asked for directions.

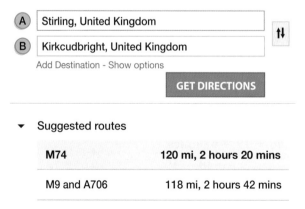

He's offered two routes.

a Using the M74, the distance is 120 miles and the estimated time is 2 hours 20 minutes.

What average speed is the computer assuming the driver can do on the M74?

b Calculate the average speed assumed if he uses the A713.

c Why might he take longer even though the A713 is a shorter distance?

7 On a motorway the maximum speed allowed is 70 mph.

Betty drove 180 miles along a motorway in two-and-a-half hours.

a Did Betty break the speed limit?

b Explain your answer.

8 James from Paisley is picking up his friend Bill from Beith to go climbing in Arran.

From Paisley to Beith is 12·6 miles. He can average 55 mph on this stretch of road.

From Beith to the Ardrossan ferry is a distance of 13·9 miles but, being a country road, he can only manage to average 40 mph.

It takes five minutes for Bill to load his stuff in the car at Beith before they get going again.

What is his average speed for the whole journey from Paisley to the Ardrossan ferry?

9 Liam leaves Glasgow and starts his drive to Inverness.

A little later Lisa starts out on the same journey.

The distance–time graph gives details of the journeys.

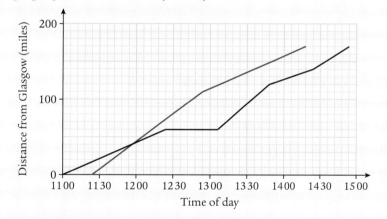

a Who arrived first in Inverness?

b When did the first person arrive in Inverness?

c How many minutes later did the second person arrive in Inverness?

d **i** Who stopped for lunch?   **ii** How long did they stop?

e **i** At what time were they the same distance from Inverness?
   ii What was this distance from Inverness?

f One of the drivers was slowed down, due to an accident.
   i Who was held back?   **ii** For how long?

g Calculate the average speed of each driver for the whole journey.

h Investigate the section of the journey that had the greatest average speed for each driver.

10 **a** Sound travels 500 metres in 1·457 seconds.

Calculate the speed of sound in: **i** m/s **ii** km/s.

b Calculate the speed of sound in mph.

(Convert from m/s to mph using 1 m/s ≈ 2·2369 mph.)

c Light travels 4 497 000 km in 15 seconds.

Calculate the speed of light in km/s.

d Why do you think you see a flash of lightning before you hear the clap of thunder?

11 The wheel-driven land speed record was broken in 2010 in Salt Flats, USA by Charles Nearburg, who reached a speed of 667·037 km/h.

Calculate the record speed in miles per hour, to the nearest whole number.
(Conversion: 1 km/h = 0·621 miles per hour.).

12.3 Calculating distance

A cyclist travels at a steady speed of 25 km/h.

In one hour he will have travelled 25 kilometres.

After three hours he will have travelled (25 × 3) kilometres = 75 kilometres.

The distance travelled is found by multiplying the speed by the time, i.e.

distance = speed × time

$$D = S \times T$$

Example 1

The train journey from Edinburgh (Waverley) to London (Kings Cross) took 5 hours 8 minutes, travelling at an average speed of 80 mph.

Calculate the distance between the two stations.

Time (in hours) = 5 hours 8 minutes = $5\frac{8}{60}$ hours = 5·1333333... hours.

The distance, $D = S \times T$, so $D = 80 \times 5.1333333... = 410.666666...$ miles.

The distance from Waverley to Kings Cross is 410·7 miles (to 1 d.p.).

Exercise 12.3A

1 Use the formula $D = S \times T$ to work out the distance travelled in:

   a 2 hours at a steady speed of 25 km/h

   b 6 hours at an average speed of 18 mph

   c 7 minutes at a constant speed of 6·5 metres per minute

   d 12 seconds at 9 m/s

   e 1 day at an average speed of 700 km/h

   f 4 minutes at a speed of 3 m/s.

2 A river flows at 2·5 metres per second.

How far will a piece of driftwood in the river travel in:

   a 30 seconds   **b** 5 minutes?

3

Cheetah 31 m/s   Lion 22 m/s   Greyhound 17·6 m/s   Human 12·5 m/s   Elephant 11·2 m/s

How much further can the cheetah travel in 30 seconds than the lion, the greyhound, the human and the elephant?

4 Tom cycles from home to school at a steady 16 kilometres per hour. The journey takes him 15 minutes. How far from the school does Tom live?

5 The *Supreme* electric scooter has a maximum speed of 15 mph.

It will travel for 30 minutes before the battery needs to be re-charged.

   a How far can the *Supreme* travel before the battery needs re-charging?

The maximum speed of the *Ultra* electric scooter is 10 mph.

Forty minutes of continuous use is guaranteed before its battery needs to be re-charged.

   b What distance can the *Ultra* cover between charges?

   c **i** Which scooter travels further?   **ii** By how much?

6 The new Golden Arrow motorbike is on a series of test-runs.

It runs for 90 minutes at 80 mph, 2 hours 15 minutes at 100 mph and for 30 minutes at 130 mph.

What is the total distance covered in the test-run?

Exercise 12.3B

1

Spider   Giant tortoise   Garden snail
0·52 m/s   0·08 m/s   0·0134 m/s

(The speeds given are the maximum speeds for each creature.)

How much further in a minute can the spider travel than:

a the giant tortoise

b the garden snail?

2 A 'domestic' tortoise can move at a speed of 0·48 km per hour.

How far can it move in five minutes? Give your answer in metres.

3 The highest speed obtained by a manned spaceship is
39 895 km/h by *Apollo 10*.

How far did it travel at that speed in:

a one minute

b one second?

4 A homing pigeon can fly at speeds of around 92 mph for up to $4\frac{1}{2}$ hours.

At that speed, what distance can it cover in $4\frac{1}{2}$ hours?

5 Sinita and her friends walked round the Isle of Cumbrae.

It took them 2 hours 12 minutes, walking at an average speed of 6 km/h.

How far is it round the island?

6 To get fit, Ann goes to the gym and uses the treadmill.

She sets its speed at 8 km/h and walks on it for 45 minutes.

a How far does Ann walk?

b Would she walk a greater or smaller distance if she increased the speed of the treadmill to
9 km/h and reduced her walking time to 40 minutes?

7 A mountain rescue helicopter left its base at 11 35
and reached the injured climber at 12 55.
Its average speed was 132 km/h.

How far did it have to fly?

8 If Shari averages 45 mph driving from Stirling to her home, she arrives in three hours.

   a How far from Stirling does Shari live?

   b Shari only manages to average 38 mph.
      How far short of home is she after driving for three hours?

9 Ronnie is training for the Olympic Games triathlon.

He swims for 18 minutes at 5 km/h.

He cycles for 2 hours at 20 km/h.

He runs for 48 minutes at 12·5 km/h.

Calculate the distances he covers at each stage.

(Remember: 18 minutes = $\frac{18}{60} = \frac{3}{10} = 0\cdot3$ of an hour.)

10 The Earth takes one year to go round the Sun.

It travels at a speed of 1·6 million miles a day.

There are 365 days in a year.

How far does the Earth travel in going once round the Sun?

Preparation for assessment

1 Tariq often works the night shift in a hospital.

  His shift lasts from 20 45 until 07 15.

  How long is Tariq's shift?

2 A power cut in the village of Auchenharvie lasted from 7.36 p.m. until 3.47 a.m. the following morning.

  For how long was Auchenharvie without electricity?

3 Millie's flight to Hong Kong left London at 22 36.
  The flight took 11 hours 35 minutes.

  When did Millie arrive in Hong Kong?

4 Sam weighed 76·30 kilograms.

He went on a diet and started taking regular exercise.

After six weeks his weight had fallen to 58·86 kilograms.

Calculate the rate at which Sam lost weight, in kg/week.

5 a Mel is paid £201·25 for working a 35-hour week in a petrol station.

Calculate his hourly rate of pay.

b At the pumps, a driver is told that in this sale he has spent £41·66, buying 29·76 litres of diesel.

The pump should tell him the rate, 'Pence per Litre'.

What is this rate, correct to 1 decimal place?

6 Ronaldo scored 57 goals in 64 games of football.

Messi scored 43 goals in 49 games.

a Which player had the higher scoring rate?
   (Calculate the goals per game.)

b How much higher?

7 In the late 1800s, a homing pigeon was released in Africa.

The pigeon took 55 days to fly home to England, a distance of 7100 miles.

Calculate the bird's average speed in:

a miles per day   b miles per hour.

8 Sandra is planning a route from Bo'ness to Lanark.

A Google Map enquiry gives:

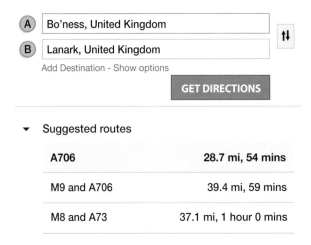

a   Calculate the average speed using:

   i   the A706 only   ii   the M9 and the A706   iii   the M8 and the A73.

b   If she went by route i and came back via route ii, what would be her average speed for the whole trip?

9   *The Pedal Pushers* cycle race is in three sections.

   Section 1: The winner averaged 34 km/h in a time of 3 hours 15 minutes.

   Section 2: The winner averaged 55 km/h in a time of 2 hours 35 minutes.

   Section 3: The winner averaged 63 km/h in a time of 2 hours 58 minutes.

a   Calculate the length of each section of the race.

b   What was the total length of the race?

c   One of the sections covered ground that was mainly flat. Another section was slightly uphill, and the remaining section was slightly downhill.

   Determine which section was:
   i   mainly flat   ii   slightly uphill   iii   slightly downhill.

10   On Monday a goods train travelled at an average speed of 52 mph for five hours.

   On Tuesday it travelled at an average speed of 44 mph for six hours.

a   On which day did the train travel the greater distance?

b   How much greater?

c   What was the train's average speed over the two journeys?

11   A tourist drove along the A841, reaching Machrie at 3.55 p.m.

   He passed through Pirnmill at 4.08 p.m.

a   What was his average speed between Machrie and Pirnmill?

b   If he had gone from the signpost to Blackwaterfoot at the same average speed, how long would it have taken him?

12   Remember Tony? He drove his car round a test circuit at a steady speed of 70 mph.

   The three gallons of petrol he had in his fuel tank lasted for $2\frac{1}{2}$ hours.

   He then repeated the exercise travelling at a steady speed of 50 mph.

   This time the three gallons of petrol lasted for $3\frac{3}{4}$ hours.

   In which trial did the car travel further?

   By how much?

Preparation for assessment

No calculator

You will be asked to do a test that requires you to perform calculations without a calculator.

The topics tested and covered will include:

- add, subtract, multiply and divide whole numbers
- rounding (significant figures and decimal places)
- find whole number percentages
- percentage increase and discount
- time, distance, speed
- direct proportion and ratio
- reading scales

- multiplication by two digits
- add and subtract negative numbers
- vulgar fractions
- time intervals
- the mean
- perimeter, area, volume
- probability.

For non-calculator work you should remember the following facts:

$$50\% = \tfrac{1}{2}, \quad 25\% = \tfrac{1}{4}, \quad 12\tfrac{1}{2}\% = \tfrac{1}{8}, \quad 20\% = \tfrac{1}{5}, \quad 10\% = \tfrac{1}{10}, \quad 33\tfrac{1}{3}\% = \tfrac{1}{3}.$$

Often calculations can be made easier by rearranging the numbers or factorising.

Example 1

Evaluate, without the aid of a calculator:

a $25 \times 38 \times 4$ **b** $14 \times 26 \times 5$ **c** 21×47 **d** 28×61

a $25 \times 38 \times 4 = 25 \times 4 \times 38$ Rearranging to bring factors of 100 together
$= 100 \times 38$
$= 3800$

b $14 \times 26 \times 5 = 14 \times 5 \times 26$ Rearranging
$= 70 \times 26$
$= 10 \times 7 \times 26$ Factorising: $\times 70$ is same as $\times 10 \times 7$
$= 260 \times 7$
$= 1820$

c $21 \times 47 = 3 \times 7 \times 47$ Factorising: $\times 21$ is same as $\times 3 \times 7$
$= 3 \times 329$
$= 987$
OR: $21 \times 47 = (20 + 1) \times 47$
$= (20 \times 47) + (1 \times 47)$
$= 940 + 47 = 987$

d $28 \times 61 = 4 \times 7 \times 61$ Factorising: $\times 28$ is same as $\times 4 \times 7$
$= 4 \times 427 = 1708$
OR: $28 \times 61 = (30 - 2) \times 61$
$= (30 \times 61) - (2 \times 61)$
$= 3 \times 610 - 122$
$= 1830 - 122 = 1708$

Example 2

A pair of trainers normally cost £60.

In a sale, they are reduced in price by 35%.

What is the sale price?

TRAINERS
35% OFF

$$10\% = £6 \qquad \left(\tfrac{1}{10} \text{ of } £60\right)$$
$$30\% = £18 \qquad (3 \text{ times } 10\%)$$
$$5\% = £3 \qquad \left(\tfrac{1}{2} \text{ of } 10\%\right)$$

35% of £60 = (30% + 5%) of £60 = £18 + £3 = £21.

So, sale price of trainers is £60 − £21 = £39.

Example 3

The National Minimum Wage for 16- and 17-year-olds in 2004 was £3·00 an hour.
By 2012 it had risen to £3·68 an hour.

Calculate the percentage increase over the eight years, giving your answer correct to 1 decimal place.

The actual increase over the eight years is £3·68 − £3·00 = £0·68.

Increase expressed as a fraction of the original: $\dfrac{0·68}{3} = 0·22666\ldots$

$$\begin{aligned}
\text{Percentage increase} &= 0·22666\ldots \times 100\% \\
&= 22·666\ldots\% \\
&= 22·7\% \text{ (to 1 d.p.)}
\end{aligned}$$

The percentage increase over eight years is 22·7% (to 1 d.p.).

Exercise Test A

1 Find the value of:

    a 26·37 + 42·86 − 38·54   **b** 184·3 − 28·76   **c** 69·45 × 7   **d** 70·08 ÷ 8.

2 Evaluate:

    a 4 × 47 × 25   **b** 5 × 24 × 8   **c** 31 × 84   **d** 19 × 45.

3 Calculate:

    a 50% of 18 kg   **b** 25% of £8·64   **c** 75% of 28 cm

    d 20% of €435   **e** 60% of 40 metres   **f** $33\tfrac{1}{3}\%$ of £14·61

    g 40% of 120 kg   **h** 15% of 750 m   **i** 45% of 950 kg.

4 Find the value of:

    a $\tfrac{1}{5}$ of 60 seconds   **b** $\tfrac{1}{8}$ of 192 kg   **c** $\tfrac{1}{6}$ of 216 litres.

5 Calculate:

    a 15% of £84   (Hint: 10% + 5%)

    b 24% of £160   (Hint: 10% + 10% + 1% + 3%)

    c 85% of 300 m   (Hint: 50% + 25% + 10%)

6 **a** Round each number to 2 decimal places:

      i 5·368   **ii** 26·082   **iii** 14·108   **iv** 3·00846.

    b Round each number to 3 significant figures:

      i 6176   **ii** 8·092   **iii** 14·862   **iv** 0·08435.

7 **a** The temperature at midnight in Aviemore was −7 °C.
      By 6 a.m. the temperature had risen by 5 °C.
      What was the temperature at 6 a.m.?

    b The temperature at Glasgow Airport at midday was 3 °C.
      By 9 p.m. it had dropped by 7 °C.
      What was the temperature at 9 p.m.?

8 George's Council Tax increased from £1220 to £1320.
Calculate the percentage increase. Give your answer correct to 2 decimal places.

9 Postage stamps can be bought in books of 8 or books of 12.
A book of 8 stamps costs £4·40. How much would you expect to pay for a book of 12 stamps?

10 The ratio of boys to girls in Third Year is 4 : 5. There are 216 students
in Third Year.
How many: **a** boys   **b** girls   are there in Third Year?

11 With his skipping rope, the boxer,
Tommy Wright-Hook, can manage
693 skips in seven minutes.
Calculate his rate of skipping in skips per minute.

12 **a** The Sun set at 8.43 p.m. and rose the
      following morning at 7.21 a.m.
      For how long was the Sun set?

    b The Moon rose in a clear sky at 21 38 and set at 05 32.
      For how long was it visible?

13 Jess took three hours to drive 132 miles from Stirling to Carlisle.
What speed did Jess average, in mph?

14 How long would it take a train, travelling at an average speed of 60 mph,
to cover a distance of 200 miles?
Give your answer in hours and minutes.

15 A cylindrical tank has a circular top of radius 10 m.
Its height is 20 m.
Given $\pi = 3.14$ and the formulae
$C = \pi D$, $A = \pi r^2$, $V = \pi r^2 h$, calculate:

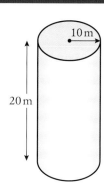

a the circumference of the top

b the area of the top

c the volume of the tank.

16 Give the readings on each of these scales as accurately as you can.

a

litres

4   5

b

g

2   3   4

c

kg

20   30

17 Sam measured the rainfall every day for a week.
The table shows his collected data.

Day	1	2	3	4	5	6	7
Rainfall (cm)	1·8	2·1	0	0·8	1·2	2·6	0·4

Calculate the mean **daily** rainfall over the week.
Give your answer in centimetres rounded to 1 decimal place.

18 Sinita and Tom have been to the supermarket. They bought:

    3 cartons of apple juice at £1·99 each
    500 g of green grapes at £3·50 per kg
    6 oranges at 49p each
    2 packs of surface wipes at £1·85 each
    white loaf at £1·45
    8 jelly-pots at 49p each
    0·25 kg loose onions at £0·92 per kg.

Calculate their total bill.

19 What is the probability of getting:

a an even number in one throw of a dice

b a prime number in one throw of a dice

c a total of eight in two throws of a dice?

20 At the last election, 9144 people voted in a constituency.

MacGregor received $\frac{1}{6}$ of the total votes cast, MacDuff received $\frac{3}{8}$ of the votes,

MacKenzie received $\frac{5}{12}$ of the votes, $\frac{1}{24}$ of the votes were spoiled.

a How many votes did each of the following candidates receive?
   i MacGregor   **ii** MacDuff   **iii** MacKenzie

b Who won the election?

c How many voting papers were spoiled?

Calculator allowed

You will also be asked to do a separate test where the use of a calculator is permitted.

You could be asked to:

- work with direct proportion
- calculate interest
- compare data sets
- use Pythagoras' theorem
- find area and volume
- plan a navigation course
- calculate probability

- interpret a wage slip
- interpret graphs and charts
- draw a line of best fit on a scatter graph
- calculate time intervals
- construct a scale drawing
- pack a container.

Exercise Test B

1 *Braised monkfish (serves 4):*

4 × 190 g monkfish fillets	*100 g shallots*
8 scallops	*60 ml white wine*
300 g mussels	*250 ml fish stock*
8 baby fennel	*120 g butter*
4 star anise	*100 ml cream*

Rewrite the ingredients for braised monkfish to serve six people.

2

Employee	Basic pay	Overtime	Bonus	Date
J. Fox	£212·36	£19·84	£25·00	30/3/13
Total gross pay	**National Insurance**	**Income tax**	**Total deductions**	**Net pay**
a	£16·14	£28·94	b	c

Jamie's pay slip is shown.

Calculate the values of **a**, **b** and **c**.

3 Melanie has £1500 to put into a savings account for two years.

> **CLYDE BANK**
> Your money grows with us
> Interest fixed for 2 years at
> **3%**
> per annum

> **FORTH BANK**
> We'll invest in your future
> for 2 years
> YEAR 1 interest rate 2%
> YEAR 2 interest rate 4%

In which bank should Melanie save her money?

Explain your answer.

4 The graph shows the mortality rates from lung cancer in the UK from 1971 to 2010.

From the graph check that in the UK in 1971 about 56 males in every 100 000 died from lung cancer.

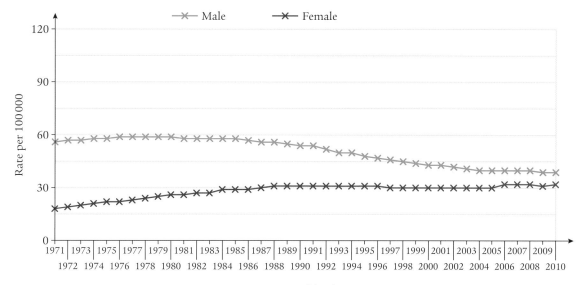

Year of death

a Estimate the number of females per 100 000 who died from lung cancer in:
  i 1971   **ii** 2010.

b Calculate the percentage increase in the number of females dying from lung cancer from 1971 to 2010.

c Calculate the percentage drop in the number of males dying from the disease over the same period.

5 The ages in years of the 15 players in both the Scotland and England rugby teams are given in the table.

Scotland team, age (years)	31	28	24	29	27	32	26	22	24	20	25	23	30	33	28
England team, age (years)	29	25	32	29	28	26	34	25	27	29	34	28	32	29	31

a For the Scotland team, calculate:
  i the range of ages   **ii** the modal age   **iii** the mean age.

b Calculate the same three pieces of information for the England team.

c Compare your answers to **a** and **b**.

6 Sixteen youngsters aged between 8 and 18 were chosen at random.

They were asked to run 60 metres as fast as they could.

Their times, to the nearest second, are given in the table:

Age (years)	14	12	15	9	8	18	16	13	15	17	9	15	17	12	17	13
Time (s)	10	12	10	14	14	9	10	11	9	9	13	10	10	9	9	12

a Draw a scatter graph to illustrate the data (put age along the horizontal axis and time up the vertical axis).

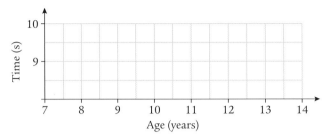

b There is one point in particular that seems out of place. Which point?

Explain what that point represents.

c Ignoring the point mentioned in **b**, draw the line of best fit.

d Describe the correlation between the age of a child and their time in running 60 metres.

7 The *Daily News* reported the following information.

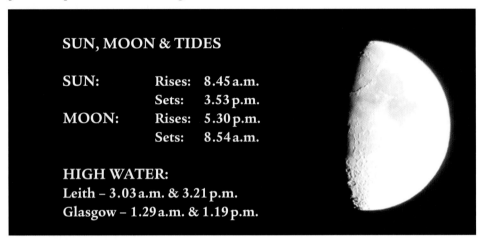

SUN, MOON & TIDES

SUN:	**Rises:**	**8.45 a.m.**
	Sets:	**3.53 p.m.**
MOON:	**Rises:**	**5.30 p.m.**
	Sets:	**8.54 a.m.**

HIGH WATER:
Leith – 3.03 a.m. & 3.21 p.m.
Glasgow – 1.29 a.m. & 1.19 p.m.

a For how long is the Sun risen?

b For how long is the Moon risen?

c How many hours and minutes are there between the high tides at:
  i Leith   **ii** Glasgow?

d Rewrite the notice with the times changed to 24-hour times.

8 Monkton Metals make tool boxes from sheet metal.

The dimensions of a tool box (a cuboid) are
80 cm by 40 cm by 60 cm.

a Sketch a net of the tool box.

b What area of sheet metal is required to make a tool box?

c Calculate the volume of the tool box.

d The company wish to make a smaller tool box, with half the volume by reducing each of its dimensions.

Suggest suitable dimensions for this new tool box.

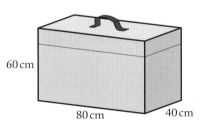

60 cm

80 cm   40 cm

Area, $A = L \times B$
Volume, $V = L \times B \times H$

9 The diagram is a rough plan of a house extension.

  a **i** Using a scale of 1 : 100, make a scale drawing of the extension. Measure the breadth of the windows properly, centring them in the middle of the wall.

  ii Mark the doors but ignore their width and place them approximately.

  b Use your scale drawing to find the actual length of diagonal AB.

  c Calculate the area of the extension in m².

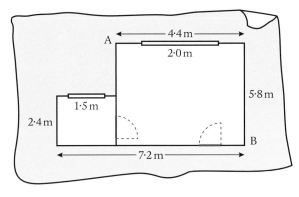

10 The *High Spirits* (H) is in trouble.

There are two boats in the area, *The Wanderer* (W) and the *Knotty Nancy* (K), which can come to the rescue. The *High Spirits* is 28 km from *The Wanderer* on a bearing of 035°.

The *Knotty Nancy* is 30 km from *The Wanderer* on a bearing of 100°.

  a Using a scale of 1 cm : 4 km, make a scale drawing.

  b From your scale drawing find:
  i which rescue boat is closer to *High Spirits*
  ii by how much.

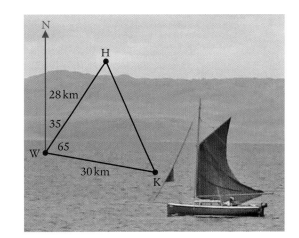

11 Cylindrical tins of tuna are packed into boxes. (In the diagram the sizes are given in centimetres.)

The boxes (cuboids) are designed to hold 36 tins of tuna.

  a Suggest one set of possible dimensions for the box (in whole numbers of centimetres).

  b Calculate the surface area of the box with your chosen dimensions.

  c Other boxes are possible. Give the dimensions of the box that will use the least cardboard.

  d Calculate the volume of a tin of tuna, to 2 decimal places.

12 Exchange rates vary from bank to bank.

SCOTIA BANK
NO commission.

Rate of Exchange:
£1 = 1·264 euros

ALBA BANK
The best rates of exchange:
£1 = €1·2708
+
1·5% commission

Avril buys £500 worth of euros from Scotia Bank.
Andy buys £500 worth of euros from Alba Bank.

a Who got the better deal?   **b** By how much?

13 A local election is held.

The share of the votes is shown in the pie chart.

There were four candidates: Liberal Democrat, Conservative, Labour and an independent.

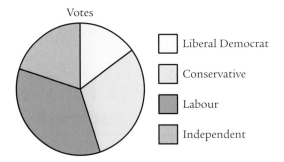

Votes

☐ Liberal Democrat

☐ Conservative

◼ Labour

◼ Independent

a What percentage of the total votes cast were obtained by:
   i the Liberal Democrat   **ii** the Labour candidate?

b If there were 20 000 votes cast, estimate the number of votes cast for:
   i the Conservative candidate   **ii** the independent candidate.

14 The floor of the community hall is in the shape of a trapezium.

a Calculate the area of the hall.

$$\left(\text{Hint: } A_{\text{trapezium}} = \frac{h}{2}\,(a + b)\right)$$

b Use Pythagoras' theorem to calculate the length of LM (to 1 d.p.).

c Calculate the length of KN.

d What is the perimeter of the hall?

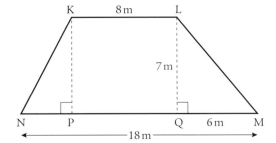

15 A pack of cards is shuffled and placed face down.

A card is picked at random.

a Calculate the probability that the card picked is:
   i a red card   **ii** a face card
   iii not a face card   **iv** a red or a black card
   v the nine of clubs   **vi** a black face card.

b A card is picked at random, noted, and returned face down.

The pack is shuffled and placed face down.

A card is picked at random, noted, and returned face down.

This is repeated until it has been done 100 times.

How many times would you expect to pick:
   i a red card
   ii a face card
   iii an ace
   iv a black face card?

(Round your answers to the nearest whole number.)

16 Three plumbers have different scales of charges.

Joe Tapp has a call-out charge of £30 and an hourly charge of £30.

Will Pipe has a call-out charge of £50 and an hourly rate of £20.

Betty Flush has a call-out charge of £65 and an hourly rate of £15.

a Joe's charges can be expressed as $C = 30x + 30$, where C is the total charge in £s and x is the number of hours he works on the job.

Write a similar equation for:
   i Will's charges
   ii Betty's charges.

b On the same graph draw the three lines representing the charges of the plumbers.

Use 2 mm graph paper and use the scales shown below.
Continue the 'Charge' axis up to £200.

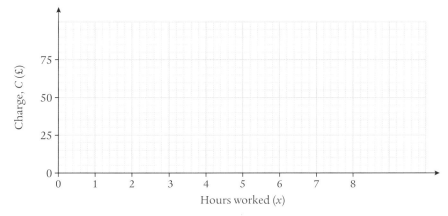

From your graph answer these questions.

c Which plumber is cheapest if the job lasts:
   i one hour   **ii** two hours   **iii** three hours?

d   **i** Which plumber would you use for a job estimated to last five hours?
   ii How much cheaper would this plumber be than the other two?

Index

Acknowledgements

The authors and publishers are grateful to the following for providing photographs:

iStockphoto: 1, 2b, 3t, 5, 6, 12, 18, 45, 53, 60, 61b, 66, 67, 69t, m, b, 71, 74, 76, 77t, b, 79, 80, 81b, 103, 122, 137, 170, 171b, 173, 174, 175t, b, 177t, b, 178, 189t, b, 198, 205t, b, 206, 210, 212, 214t, 215t, b, 216, 217t, b, 218, 219, 221, 230, 232, 234, 235 (all), 236 (all), 237b, 239, 242m

www.victorianweb.org: 164

All other photographs were provided by the authors.

Microsoft product screenshots reprinted with permission from Microsoft Corporation.

Microsoft and its products are registered trademarks of Microsoft Corporation in the United States and/or other countries.

254